Get Funded: An Insider's Guide to Building An Academic Research Program

Learn all the basic principles involved in initiating an academic career and building an externally funded academic research program with this practical guide. Based on the author's extensive experience as a government funding agency director and successful academic, it provides step-by-step advice on how to identify an appropriate funding agency and program manager, how to present your research in a concise and effective manner, and, ultimately, how to obtain your first research grant. It explains the faculty recruitment process in detail and outlines the key timelines associated with being on the tenure track. Providing a unique insight into research funding agency operation and expectations, this is the "go to" guide for new faculty members in engineering, the sciences, and mathematics looking to gain a head start in their academic careers.

ROBERT J. TREW is the Lancaster Distinguished Professor in the Department of Electrical and Computer Engineering at North Carolina State University (NCSU). He has previously served as the ECE Department Head at NCSU, Virginia Tech, and Case Western Reserve University, as a Program Manager for the US Army Research Office, as the Director of the ECCS Division of the National Science Foundation, and as the Director of Research for the US Department of Defense. He is a Fellow of the AAAS and a Life Fellow of the IEEE.

Get Funded: An Insider's Guide to Building An Academic Research Program

Robert J. Trew

North Carolina State University

CAMBRIDGE
UNIVERSITY PRESS

CAMBRIDGE
UNIVERSITY PRESS

University Printing House, Cambridge CB2 8BS, United Kingdom

One Liberty Plaza, 20th Floor, New York, NY 10006, USA

477 Williamstown Road, Port Melbourne, VIC 3207, Australia

314-321, 3rd Floor, Plot 3, Splendor Forum, Jasola District Centre, New Delhi - 110025, India

79 Anson Road, #06-04/06, Singapore 079906

Cambridge University Press is part of the University of Cambridge.

It furthers the University's mission by disseminating knowledge in the pursuit of education, learning and research at the highest international levels of excellence.

www.cambridge.org
Information on this title: www.cambridge.org/9781107657199

© Cambridge University Press 2017

First published 2017

A catalogue record for this publication is available from the British Library

Library of Congress Cataloging in Publication data
Names: Trew, Robert J., author.
Title: Get funded : an insider's guide to building an academic research program / Robert J. Trew.
Description: New York : Cambridge University Press, 2017.
Identifiers: LCCN 2016057390| ISBN 9781107068322 (hardback) | ISBN 9781107657199 (paperback)
Subjects: LCSH: Proposal writing for grants – United States. | Federal aid to higher education – United States. | Research grants – United States.
Classification: LCC HG177.5.U6 T74 2017 | DDC 001.4068/1–dc23
LC record available at https://lccn.loc.gov/2016057390

ISBN 978-1-107-06832-2 Hardback
ISBN 978-1-107-65719-9 Paperback

To my family, Diane, Heather, and Robin, for their continuing encouragement and support.

Contents

Preface

The material presented in this book is directed towards helping someone either searching for, or just embarking upon, an academic faculty position, to learn how to go about establishing an externally funded research program. The search for research funding is a never-ending struggle, and one that is becoming increasingly competitive. Getting a head start by learning how research grant funding agencies operate, and learning what program officers look for in a new researcher, can facilitate the process. This book is dedicated to providing this information. However, before proceeding, and to give you some confidence that you are not wasting your time by reading this book, let me say a few words about myself, specifically my credentials regarding my knowledge relative to establishing and funding an academic research program, and why I am writing this book.

First, I have had an extensive and diverse career, and have worked over the past four decades in a variety of industrial, government, and academic organizations. Basically, I have experience, and have served on all sides of the academic and research program enterprise, extending from starting my own academic career, as well as serving as an academic department head involved in recruiting and mentoring young faculty, to initiating and managing funded research programs as a US government research program manager and director. I served for about eight years as a Program Manager in the Electronics Division of the US Army Research Office, where I was actively engaged in both the identification and definition of new research areas, and the establishment and management of directed research program funding opportunities. I also served for over three years as the Director of Research for the US Department of Defense (DOD), with management oversight responsibility for the entire basic research program that is sponsored and funded by DOD. Also, I served for four years as Director of the Electrical, Communications, and Cyber Systems Division in the Engineering Directorate of the National Science Foundation.

Altogether, I have about 15 years' experience working in various program director and research program management positions within US government research funding agencies. In my US government research manager and director role I worked within US government research funding agencies, collaborated with program managers in numerous and various US government funding agencies, and helped to identify and define research areas for future sponsored research, and then helped to define specific topics for directed research program funding. I have been, and am currently, also heavily involved in the review and evaluation of research programs, both external and internal to US government agencies. As a US government program manager, I initiated and managed academic research programs to address specific research topics that were directed to nationally identified research problems. In this role I have had significant experience working with both young and experienced faculty members working on academic research projects. Over my career I have personally read and evaluated a very large number of research proposals, numbering in the high hundreds and approaching 1000, and I am well aware of the elements of a good research proposal, as well as pitfalls that can result in a poorly written proposal that has little chance of obtaining funding. Over time, and with experience, I have learned how to read a research proposal and extract the significant attributes in a very efficient manner and with minimal time expenditure. An excellent research proposal should be written very concisely and effectively so that the reader can quickly and efficiently learn what is being proposed.

Second, I am fundamentally an academic engineer, although I have worked in industry, major universities, and the US government, with a career extending over essentially four decades, heavily focused upon research. I have built a successful academic research program as a university faculty member, and have been successful in obtaining research funding adequate to support a significant number of graduate students and my research activities. In this effort I have probably written close to 100 research proposals, with a high rate of success in obtaining funding. I have, to date, served as mentor and faculty advisor to a significant number of PhD and MS students. All have been funded from Research Assistantships from research funds that I have obtained from US government funding agencies and industrial research and development (R&D) grants. In addition, I have served as the Head of the Electrical and Computer Engineering Departments at three major research-oriented universities: North Carolina State University, Virginia Tech, and Case Western Reserve University. In this academic department management role, I have been responsible for recruiting new faculty members, and for helping them to establish research programs. In addition, I have been responsible for academic research laboratory facilities, and associated support issues. I have also had overall responsibility for recruitment of new graduate students. I am well aware of the issues that new faculty members face, and am familiar with providing advice and guidance to them.

Third, my reasons for writing this book are essentially explained above. I have gained much experience and knowledge in my career relative to building an academic research program. As stated, I have served virtually on all sides of the academic research program enterprise. I know how US government research funding agencies identify and define new research program areas, and I know what US government research funding agencies are looking for in new researchers. Also, I understand the issues associated with a new faculty member's expectations and responsibilities as they initiate an externally funded research program. My insight into the factors and problems that a new faculty member will encounter is probably unique, owing to my extensive experience in both the world of the US government funding agencies, as well as my personal experience in the academic research world. My goal is to convey as much of this information as possible to new and interested faculty members, and to help them to initiate their own research programs with as much efficiency and minimal effort as possible. I want to help new faculty members understand how the research grant funding system functions, and how they can optimize their interactions and participation. Hopefully, this information will facilitate the transition into an academic position. This is my main reason for writing this book.

1 Introduction

Are you interested in an academic career? Would you like to be a professor in a college or university, and particularly a research-oriented college or university? If so, have you wondered how colleges and universities recruit and select their faculty members, or what is required and expected of a faculty member once they have been hired? Also, once embarked upon an academic career path, have you wondered what is expected of a new faculty member in his or her research activities, and how a new faculty member initiates and establishes a research program? Do you know who pays for research that is performed in academic institutions, and how a faculty member goes about obtaining research funding? If your response to these questions is positive and the questions pique your interest, and you would like to obtain more information relevant to establishing an academic career, you have selected the correct book to read! These, and other questions are directly addressed in the following chapters in this book. We'll define our terms, and then go over the different ways academic research attracts funding. We'll explore the differences between public and private financing, and discuss the benefits and drawbacks of different funding vehicles. Hopefully, you'll find some answers that address your personal situation and shed some light onto the procedure and effort needed to establish an academic career.

As you probably know, the life of a university professor can be very rewarding and fulfilling, which may be, at least partly, why you are interested in an academic career. You may have observed faculty members in the performance of their duties and their activities in a college or university you attended, and have been inspired and motivated to pursue a similar lifestyle. You may have pursued a PhD degree with the specific intention of becoming a university faculty member. If so, you are to be congratulated and encouraged, as an academic faculty career is a noble and worthwhile endeavor, and you will be contributing to the advancement and welfare of civilization, and the education and training of the next generation.

However, various factors associated with an academic career can be initially confusing, and you probably have some questions regarding how one goes about obtaining and initiating an academic career, and what is expected from a new faculty member. You may also know that an academic career path differs substantially from an industrial or business career path in several significant respects. For example, as a business or industrial employee, particularly in technical disciplines such as scientist or engineer, your job duties and responsibilities will be defined for you, usually through discussions with your supervisor, and you will be, at least initially, working on projects that are well defined for you. As your career advances you will gain more flexibility to define projects, but this ability will always be limited by the requirements of your employer and their needs and interests. In general, you will not necessarily have the flexibility to pursue problems that you personally find interesting, particularly if they don't have immediate relevancy to your employer's goals. You will be paid at a prescribed and determined rate, depending upon your level of experience and job performance. Your performance expectations will be fairly well defined and documented, and your salary will be adjusted periodically based upon your project success and your performance.

An academic career has similar characteristics, primarily for your teaching and committee service duties. However, the situation is dramatically different regarding your research program establishment and development. The expectations for your research activities can be, and usually are, confusing, particularly for new faculty members. Establishing a funded research program effort is very much like starting your own business. In fact, you will basically be an entrepreneur. That is, you personally will be responsible for the identification and definition of the specific research area and topics that you choose to pursue. You will essentially have complete flexibility to pursue any research topic you desire. This is the good news. The bad news is that you will be expected to obtain your own funding to actually perform the research. Your home institution will provide very little financial support for your research, other than items included when you are initially recruited and hired. Of course, you will also be expected to teach courses, and to serve on academic college and department committees. This is a required obligation, and an expected responsibility of all academic faculty members, although the effort represents a demand upon your time. Unfortunately, these duties often conflict with research program activities. Obtaining funds to support a research effort is a high-priority activity, although extremely challenging, and an activity that will dominate your entire career as a university faculty member. Effective time management is a definite necessity and characteristic of all successful faculty members.

If you've decided to pursue an academic position, congratulations on the beginnings of an exciting and rewarding career. I sincerely hope you find the material in this book informative and useful. If you're reading this book you may have recently graduated with your PhD and are either looking for, or have found, your dream job in academia as a university professor. You may also have graduated in the recent past and have been employed in industry or government, or possibly have worked as a postdoctoral research assistant in an academic research group, and recently began the faculty recruitment process, or perhaps you have accepted a university faculty position. Or perhaps, you've been in academia for a while and are now looking to change your research direction.

Whatever your path to an academic position, you've been involved in and survived, a very intense and competitive process. Securing an academic tenure track position is one of the most competitive processes you'll ever encounter. If you've been recruited to an academic faculty position, you likely have been selected over a large number of competitors, which could number in the hundreds. As a university department head I've been involved in numerous new faculty searches where we've received on the order of up to 100 or more applicants for a single position. Academic faculty opportunities in US academic institutions now typically will attract tens to hundreds of applicants, and from candidates from all over the world. The competition for academic faculty opportunities has dramatically changed over the past three or four decades. Back in the 1970s and 1980s, for example, many new faculty opportunities at US academic institutions often experienced problems recruiting faculty members. Although there were numerous reasons for this, faculty recruitment difficulties were, at least, partly due to relatively low academic faculty salaries that were paid at the time. Academic salaries lagged significantly behind industrial salaries. However, the situation has leveled in recent decades, and academic faculty salaries, particularly in engineering and science disciplines, but also in disciplines such as business, marketing, communications, etc., now have relative parity with industrial and government pay scales. Academic faculty positions are very highly desired, and new faculty searches at most major universities have become very competitive in recent years. If you've secured a tenure-track academic faculty position, you are to be congratulated! You've survived a very intense process and have positioned yourself for a life in academia, which, quite frankly, can be a very enjoyable and satisfying lifestyle, while contributing to further discovery and passing this knowledge on to the next generation.

While you bask in your current success, you need to focus upon the next step in the process, the one for which you were primarily hired. Unfortunately, this is also the challenge for which you're probably the least prepared. As a PhD student you've been mentored and worked under the direction of your advisor, who may have had the responsibility of securing your funding while you worked on your research on projects for which your advisor wrote a successful proposal and received funding. You're now faced with the goal of building a successful research program on your own. You'll find that moving from one side of the desk to the other can be daunting, particularly since you probably don't yet have a lot of contacts in funding agencies and, in fact, may not even be aware of where to look. You're faced with the challenge of obtaining funding to support your research, to recruit students, and, of course, to teach classes, which, if you're lucky, will involve developing courses that support your research area. You'll find that you have an incredible amount of freedom to basically pursue whatever you want to pursue, but you also have the responsibility to obtain the necessary funding. When young faculty members fail, it's generally because of their inability to properly manage the freedom and flexibility afforded by an academic career. It's a very time-intensive process to identify and make the connections with funding agencies and the appropriate program managers that will support your research. It takes a lot of time to decide which research topic to pursue and determine how to write a proposal that will be successful in being funded.

During your recruitment process you've most likely been wined and dined and promised all sorts of help, mentoring, and advice. Most of this has been promised in a spirit of helpfulness and understanding of the challenge you face. After all, the people that just hired you have also gone through the process and survived. They know the difficulties you face. An increasing number of universities offer detailed mentoring programs that will help you become aware of funding opportunities, and many now offer proposal-writing workshops. By all means, attend as many of these as possible, for it's not possible to get too much advice. However, be aware that most of these mentoring workshops are presented from the university perspective. They may include topics such as identifying potential sponsors, how to write an effective proposal, university procedures and policies, pre-award and post-award procedures, etc. While this information is pertinent and of much use, it doesn't address the more important question of what a program manager will find worth funding. In an era of reduced research funding and an expanding base of faculty members seeking research funding, the competition for research funding has become more and more competitive. Funding rates at most funding agencies have been on a continuous decline over the past 20 years or so, and generally are no larger than 20% to 25% or less for most opportunities. In order to be successful in this environment, it's increasingly important to "tune" your proposal to the topics in which program managers have interests. Therefore, you need to learn what they're looking for and what they want to support. They have limited available research funds, particularly for new program starts, and are very selective in which new programs they choose to fund. You want to be one of them!

The purpose of this book is to help you in your quest. The book is written from the perspective of an experienced government program manager, and the book presents inside information on how to make contact with an appropriate program manager or program director, and to understand what they are looking for in proposals that are submitted to them. You need to learn and understand how to identify the research areas and projects that they support, and how you can fit into their program. Although the elements of proposal writing are discussed in Chapter 7, the main thrust of this book is not to specifically address the mechanics of how to write a proposal; this type of information is readily available in numerous other sources, but rather, to address the more important question of what subject to address in the proposal, and how to direct and "tune" your proposal to a funding opportunity, as well as how to get your proposal in the hands of a receptive agency and the appropriate program manager. This is the first step in building a successful research program and will set the course for future success. This type of information is not readily available elsewhere.

In the following chapters in this book, we'll address both the issues associated with the search for an academic faculty position, and the basic principles of building an academic research program. We'll discuss how to go about setting up an academic research program and what to ask for and negotiate in a start-up package. We'll then discuss the sources of research funding and how to identify the appropriate funding agency for your research, as well as how to go about making contact with the proper program manager. I'll describe what program managers and program directors in US government funding agencies are looking for in new researchers, and how you go about shaping your research to fit into their programs. Hopefully, this information will give you a glimpse into the thinking of program managers.

Chapter 2 starts with a brief overview of research funding history in the United States. We'll discuss the reasons the US government provides research funds to academic institutions, and why a variety of funding agencies exist. We'll also discuss funding trends over time, and indicate some emerging areas of research. Chapter 3 digresses from the question of how to work with funding agencies to discuss the academic faculty search process, what is expected of a new faculty member, and how to negotiate an appropriate start-up package that includes university supplied resources that will be required to initiate a research program. This information is primarily intended for those just starting to search for a faculty position, and the seeds for a successful academic career are often sown in the process of negotiating the resources that will be provided at the beginning of academic employment. New or established faculty members that have already accepted or hold a faculty position can skip this chapter, and proceed to the rest of the book, which is directed towards describing the procedures involved in obtaining the funding necessary to build and sustain an academic research program.

If one is to be successful in raising research funds, there is significant salesmanship involved. This issue is addressed in Chapter 4, where we discuss the best approach to presenting your ideas to a program manager and how to convince them that they should support your research. Chapter 5 discusses the issue of how to identify the most appropriate grant funding agency to approach for research funding. The various US government funding agencies, and the areas of research they support, are discussed. Also, the legal instruments that are used to transfer research funds from the funding agencies to research performers are explained and defined. The very important question of how to identify and make contact with the appropriate program manager or program director that may be interested in funding your research is addressed in Chapter 6. The best and most effective means for making contact with program managers and program directors are presented and discussed. Suggestions are offered regarding both how to identify and communicate with an appropriate program manager or program director, as well as how best to approach them, and learn what research topics they have the most interest in supporting. Also, the best approaches to gain their interest in your research are discussed. I'll give you some tips as to how to sell yourself.

The elements of a well-written proposal are discussed in Chapter 7. I'll explain the process by which you can elicit feedback regarding your proposal, and how you can work with your program manager to revise and refine your initial ideas. The NSF PECASE program proposal has some fundamental requirements that differ from a standard research grant proposal. For this reason, the PECASE proposal is separately described in some depth. Chapter 7 ends with a discussion of what to do if your proposal is declined, and how you can learn from the process to improve your next proposal. The book concludes with a brief discussion of some cautions, concerns, and restrictions that are associated with performing research on certain topics and information identified by the US government as "export controlled," which requires that certain actions be taken to limit access by foreign individuals or organizations. In particular, the items and information that are identified as sensitive may be restricted under the ITAR and EAR restrictions, which are explained in Chapter 8. This may also include some restrictions associated with working with international students, and in traveling to present the results of your research in international conferences, workshops, and meetings with foreign individuals and organizations. We'll define these regulations and explain their meaning and how they affect your research.

I sincerely hope you find the material and information in this book useful. As I stated, the information is derived from many years' experience working on virtually all sides of the academic research enterprise. I hope the information in this book represents the best course you never took in grad school.

2   A Brief History of Research Funding in the United States

It is common knowledge that research costs money, but who provides the funds used to support research? The answer is complex, and there are a variety of organizations that provide support for academic research, with the vast majority of research support funds that are available to academic researchers provided by the US government, state and local governments, business and industrial organizations, and a variety of charitable and not-for-profit foundations dedicated to either general research support or specific topics of interest. However, the single largest provider of funds for support of research to university-based researchers is the US government and its various research agencies and offices (described in Chapter 6). Annual research budgets for these organizations vary, and issues associated with research funding are routinely reported in television and radio news reports, and articles on the subject regularly appear in newspapers, magazines, and other popular press media. Various organizations, such as the American Institute of Physics (AIP), the American Association for Engineering Education (ASEE), the American Association for the Advancement of Science (AAAS), and others regularly publish bulletins that report latest developments in government agency annual budget negotiations and appropriations, particularly as they affect the academic research community. It is not difficult to find a report in the news media directed towards various issues associated with current trends in research funding, and the status of research funding budget requests from various government research agencies and the related negotiations that are in progress at any particular time. Funding for research generally, although not always, has strong popular support, and many politicians commonly advocate for an increase in research funding for various US government research funding agencies, using public funds, particularly during election years.

Also, as both national and state governments seek to develop their economies and to create industrial, business, and other employment opportunities, national and state leaders generally look to academic institutions to serve as "incubators" of new business opportunities, with the specific goal of job creation and an increase in local employment. Many programs directed to economic development have been conceived, designed, and introduced at both the national and state levels, and most of these programs provide "seed" type grants in order to help entrepreneurs establish their companies. While the national programs, such as the Small-Business Innovative Research (SBIR) program, are applied on a nationally competitive basis, many states have developed their own programs to encourage development within their state borders. All these programs have in common the focus upon both industrial and academic research and the desire to translate basic science and technology research into economic development and the creation of new business opportunities, with associated job creation and employment growth. Most of these programs attempt to combine industrial engineers and scientists engaged in R&D activities with academic researchers in order to facilitate collaboration and coordination, and to transition basic research to practical applied research that can result in products being introduced into the marketplace. This process is widespread throughout the United States, and is also commonly employed internationally.

Before discussing the academic research process and the issues associated with establishing a funded academic research program in more detail, and to gain a better understanding of the role of R&D in academic institutions, it's helpful to consider how government agencies became involved in funding research, and the overall goal of government supported research. A brief review of the history of how US government research funding agencies were established helps a researcher to understand how these agencies function, some of the issues and pressures they face, and this, in turn, helps the researcher learn how to work with the various agencies.

2.1 US Government Support for Academic Research

It may come as a surprise to some that the US government wasn't always a major supporter of academic research and, in fact, that government funding for research performed in academic institutions is a relatively new concept introduced in the twentieth century. So, how did the US government come to be the major supporter of funding for academic research?

The US government has always provided research support for certain activities, stemming back to the founding of the country. However, most of the financial support was for special projects and activities, and essentially directed towards private companies and business. Very little, if any, financial support was provided to academic scientists. How the US government became a major provider of academic research funding stems back to the country's experience during the Second World War, and the relationship that developed between the US government and academic scientists and engineers during this time. Before the Second World War, funding for research and development was very limited, significantly increasing only since the end of the Second World War. Governments have traditionally invested in technology projects, but before the Second World War, overall US government spending for research and development activities was small. In the pre-Depression years in the USA (in the 1920s), there was very little US government support provided for research in US academic institutions.

An interesting story concerning pre-war research funding relates to Robert Goddard, the father of modern rocketry [1]. Goddard was a professor of physics at Clark University in Massachusetts. In 1930, he established a laboratory in the New Mexico desert, primarily with his own funding, to research and develop liquid-fuel rockets. He picked the New Mexico desert for his rocket work because he could test his rockets far away from any buildings or people that could be hurt in case of accidents, and he could perform his experiments in relative isolation and away from any observers. Although he did secure small grants from the Smithsonian Institution and the Carnegie Foundation to support his research, and later in the 1930s larger grants from the Guggenheim Foundation, he had little luck in obtaining grants from the US government and most of his funding came from his own financial resources. Although the US Army was interested in his work, the US government declined to provide him any research support in the time period between the two world wars. However, German engineers, working on the V-2 and other rockets for the German military during the Second World War, studied Goddard's publications and patents and applied his ideas in their development efforts with good success. Later, in the 1950s, and after declining to fund his research, the US government infringed a number of Goddard's patents on liquid-fuel propulsion and gyroscopic stabilization. After a lengthy nine-year court battle Goddard's widow (Goddard died in 1945) and the Guggenheim Foundation, which had an equity stake in his Intellectual Property (IP) due to the funding they provided, prevailed in court and the US government settled with a sizeable monetary settlement [2]. At the time it was the largest monetary settlement made by the US government for patent infringement.

Funding for research activities was very limited throughout the 1800s and into the early 1900s. Universities expected faculty members to engage in research as part of their duties, but there was no clear means of obtaining funding to support the research. Herbert Hoover noted this situation, and while serving as US Secretary of Commerce from 1926 to 1930, he campaigned to establish a National Research Endowment [3], supported by funds provided by industry. Hoover warned that the United States had, up to this point, depended upon three sources for all the support of pure science research over the years: (1) that the rest of the world would bear this burden of fundamental discovery for us; (2) that universities would carry it as a by-product of education; and (3) that men of great benevolence would occasionally endow a Smithsonian or a Carnegie or a Rockefeller Institute. He felt that the future welfare of the country depended upon scientific discovery, which required a stable source of funding for university-based research. However, industry failed to see how they would reap any benefits and declined to provide support funds. The National Research Endowment never successfully secured adequate funding, and it disappeared during the Depression years that followed.

Although there were examples of US government support for academic research, there was opposition to this from an unexpected quarter. In 1935, physicist Karl Compton was president of MIT and head of the Science Advisory Board, which, in the wake of the Depression resulting from the stock market collapse in 1929, was established by executive order to address (1) the ills of unemployed scientists and (2) unmet social problems. Although a supporter of research, Compton argued in an article in *Science* [4], "If government financial support should carry with it government control of research programs or research workers, or if it should lead to political influence or lobbying for the distribution of funds, or if any consideration should dictate the administration of funds other than the inherent worth of a project or the capabilities of a scientist, or if the funds should fluctuate considerably in amount with the political fortunes of an administration or varying ideas of Congress, then government support would probably do more harm than good ...." These arguments were overcome by the events taking place in Europe in the middle to late 1930s, and the formal involvement of the United States in the Second World War on December 8, 1941.

During the war years a very successful collaboration between government and academic researchers and engineers was established. The contribution of academic scientists and engineers to national security, as evidenced in the nuclear work performed during the Manhattan Project and the radar development performed at the MIT Radiation Laboratory, demonstrated the benefit to the country of government and university collaboration and the advantages of government support of academic research. After the war, the US government sought to define, for the post-war era, the role of university scientists for peacetime and national security purposes. US President Roosevelt asked his Science Advisor, Vannevar Bush, to study the issue, and in 1945 Bush delivered his seminal work *Science: The Endless Frontier* [5] to then US President

Truman. In this work, and in response to the pre-war arguments against federal support of academic research, Bush argued for federal support of "unfettered" basic research where scientists were permitted to pursue their own ideas, and for the creation of a self-governing National Research Foundation (NRF) with divisions of medical research, natural sciences, and national defense. He also proposed a linear model for research, consisting of basic research, applied research, and advanced development. The self-governing aspect of his proposal caused significant controversy, and President Truman felt that the Constitution did not permit delegation of control over any portion of the federal budget. The NRF was never established.

Dating back to the founding of the country, government was interested in military developments stemming from research. The US Navy, in particular, saw great advantage in advancing technology for ships. However, it wasn't until the 1900s that serious efforts to engage science and advanced technology research emerged. Inspiration was provided by the inventor Thomas A. Edison, who in an interview published in the *New York Times* in May, 1915, argued that the United States, in order to prepare for the First World War, should engage with "industry and science," and that the government should create a research laboratory to work on military science projects. Secretary of the Navy Josephus Daniels read the interview and contacted Edison in order to enlist his help in establishing a research and development laboratory within the US Navy. Edison agreed to assist, and worked with Daniels to help establish the Naval Consulting Board, composed of 22 representatives from major national engineering societies to enlist private scientists and engineers to work on naval projects. The Naval Consulting Board was intended to be an interface between the US Navy and private scientists and engineers and to enlist their assistance on Navy technical problems. The Naval Consulting Board was not completely successful, but it did produce a plan to establish a research laboratory within the Navy, which Daniels successfully proposed to the administration, and was funded by Congress in 1916 with an original budget of about $1.5 million. After some delays and difficulties, the Naval Research Laboratory (NRL), located on the Potomac River in Washington, DC, was completed in

1923. It was the first research laboratory established within the US Department of Defense and employed a staff of scientists and engineers working primarily on technology to detect submarines, as well as technology to improve radios for communication. The NRL, at this time, was not well connected to the academic community.

This situation changed in August 1946 when Congress passed Public Law 588 that established the Office of Naval Research (ONR). The ONR evolved from the Office of Research and Innovation (ORI) that was established in May 1945 by Secretary of the Navy James Forrestal in order to combine several different wartime research offices with the Naval Research Laboratory. Vice Admiral Harold Bowen, who had headed the ORI, became the first Chief of the Office of Naval Research. A major goal of the ONR was to provide support to both industrial and academic researchers working on "advanced research in nuclear physics and other topics of interest to the Navy." Originally, financial support was provided for basic research projects by means of contracts. However, in 1959 support by means of grants began, and support for applied research projects began in 1980. Support for advanced technology development projects became available in 1993. The pre-war concerns voiced by Karl Compton were recognized and the ONR was organized to counter the fears that government sponsorship would be restrictive, burdened with bureaucratic rules, or subject to political pressures. Scientists were encouraged to propose their own projects. No progress reports were required and refereed publication in the open literature was sufficient evidence of progress. Support funds were made available for graduate assistants and summer faculty salary support. Awards were multi-year and renewable. The linear model of basic research, applied research, and advanced technology development proposed by Vannevar Bush was adopted and, for the most part, is still in effect today. The research category classification has since been formalized, and the latest revision occurred in 1998 when the US government issued OMB Circular A-11. In this document, basic research is defined as: "Systematic study directed toward greater knowledge or understanding of the fundamental aspects of phenomena and of observable facts without specific applications toward processes or products in mind;"

applied research is defined as "Systematic study to gain knowledge or understanding necessary to determine the means by which a recognized and specific need may be met;" and development is defined as the "Systematic application of knowledge toward the production of useful materials, devices, and systems or methods, including design, development, and improvement of prototypes and new processes to meet specific requirements."

Starting in 1959, research support funds were provided to academic institutions primarily through a grant mechanism, which does not generally carry "deliverable" requirements. This model is still, although not always, used today. Progress and final reports and program reviews have become increasingly important, and today these reports and reviews are almost always indicated as requirements for demonstration of project progress and results. Publication in refereed journals is still expected and publication is an indication of successful progress. The "unfettered" characteristic of basic research has changed considerably over the years and now, many funding opportunities tend to be more "directed" than "unfettered." There have been many studies and much written over this issue and the balance between the two approaches is continually shifting.

The other armed services soon established their own research offices, and the Army Research Office (ARO) was established in 1951, and the predecessor of the Air Force Office of Scientific Research (AFOSR) was established in 1952. In order to maintain a close relationship with the academic community, the ARO was originally located on the Duke University campus in Durham, NC. It later moved to an office in Research Triangle Park, NC, where it still resides. The other tri-service research offices, ONR and AFOSR, reside in the Washington, DC, area in Arlington, VA. In response to the Soviet launch of *Sputnik*, the Defense Advanced Research Projects Agency (DARPA) was established in 1958 to focus research development activity upon high payoff projects of interest to national security. Following the nuclear work conducted during the Manhattan Project during the Second World War, there was a need to continue nuclear oversight and this effort was managed by a series of government agencies, starting with the Armed Forces Special

Weapons Project (AFSWP) from 1947 to 1959; the Defense Atomic Support Agency (DASA), from 1959 to 1971; and the Defense Nuclear Agency (DNA), from 1971 to 1997. All of the previous efforts were moved into the Defense Threat Reduction Agency (DTRA), which was created by the 1997 Defense Reform Initiative. The charter of the DTRA was expanded to include chemical and biological research, as well as nuclear activities. The DTRA extended its research efforts to include a basic research program, primarily to support research performed in academic institutions, in 2006.

The Department of Defense (DOD), for the most part, follows the linear funding model outlined by Vannevar Bush, and provides funding for research and development in three major categories: (1) basic research (indicated as 6.1 after the budget line in the DOD budget); (2) applied research (indicated as 6.2); and (3) advanced development (indicated as 6.3). The US Department of Defense (DOD), as indicated in Volume 2B, Chapter 5, in DOD Financial Management Regulation uses the official US government definitions for basic research and applied research, but modifies the development category to advanced development, with the definition that the category "Includes all efforts that have moved into the development and integration of hardware for field experiments and tests."

The DOD budget for Science and Technology (S&T) consists of the basic research (6.1) and applied research (6.2) budget categories. Essentially all of the research funds that are provided to academic researchers originate from the S&T budget, and the DOD had a budget for Science and Technology of $12.3 billion for FY2016, including $2.1 billion for basic research and $4.7 billion for applied research.

Although the National Research Foundation (NRF) was never established, the debate that followed resulted in a civilian agency, the National Science Foundation (NSF), being authorized as an independent federal agency by Congress in the NSF Act of 1950. Public Law 81–507 was signed by President Truman on May 10, 1950, officially establishing the NSF, and the NSF became operational in 1951, with the first grants awarded in 1952. The stated goals of the NSF were "to promote the progress of science," to "advance national health, prosperity, and welfare," and to "secure the national defense." The NSF was established to support fundamental research and education across all fields of science and engineering and to help build research infrastructure and to build the nation's scientific and engineering workforce. A unique aspect of the NSF was the focus upon education, particularly science, engineering, and mathematics education, with a goal towards training the next generation of scientists, engineers, and mathematicians. The NSF budget and number of grants has risen over time, with 28 grants valued at $3.5 million awarded in 1952, the first year of operation, to the present time with almost 10 000 grants worth about $7.7 billion awarded in 2016.

The National Aeronautics and Space Administration (NASA) was created on July 29, 1958 by President Eisenhower and NASA has provided funds to support basic and applied research in a variety of science and engineering fields from the very start. NASA originated from the National Advisory Committee on Aeronautics (NACA), which had been researching flight technology for over 40 years. Along with the space program that resulted in the Moon landing under Project Apollo, NASA also conducted scientific and engineering research and worked on developing applications for space technology, particularly in weather and communications satellites. After the Moon landing, NASA developed and launched the Space Shuttle as a reusable space vehicle. The Space Shuttle flew over 130 flights before being retired in 2011. Another large space project was established in 2000 with a major effort led by the United States and Russia to build a permanent human presence in space by means of the International Space Station. This project involved the work of 16 nations. Planetary work has continued with the Mars Rover and other projects. In addition to the space work, NASA also funds research on improved safety aircraft travel, as well as improved efficiency aircraft designs and more environmentally friendly technology. The relationship between NASA and academic researchers has historically been very strong. The overall NASA budget for 2016 was about $18.5 billion, with $5.2 billion for science and $571 million for aeronautics research.

The Department of Energy also provides support for research and development, with an emphasis in the field of electrical engineering, nuclear engineering, and other disciplines associated with energy research. This support originates back to 1946 with the Manhattan Project and the development of nuclear technology. The Manhattan Project was conducted under management of the US Department of War and the Army Corps of Engineers. The laboratories that were established were the origin of the current US National Laboratories. After the Second World War there was a strategic need to continue and manage the nation's scientific capabilities and, in particular, the nuclear technology that had been successfully developed. In 1946, the Atomic Energy Act was passed and responsibility for nuclear research and development was moved from the War Department to a new independent civilian agency, the Atomic Energy Commission (AEC). The AEC operated under the direction of five Commissioners appointed by the President. In response to the oil shortage problems in the late 1960s and concerns over rising petroleum imports, Congress and President Nixon expanded the AEC research charter to include non-nuclear forms of energy and related technologies. In 1974, the AEC was terminated and the research portfolio was transferred to a newly created Energy Research and Development Administration (ERDA). The research portfolio of the ERDA was broad and consisted of nuclear research and technology, and also what we now call "alternate" energy technologies, including solar, fossil, and geothermal programs, as well as conservation, synthetic fuels, and power-transmission research. Three years later, the ERDA was absorbed, along with about 30 other energy-related functions, with the establishment of the Department of Energy, which gathered government energy related research, policy, and regulatory functions into one agency. The Energy Organization Act of 1977 created the Office of Energy Research (renamed the Office of Science in 1998), which was organized into two divisions: (1) High Energy and Nuclear Physics, and (2) Basic Energy Sciences. The Basic Energy Sciences (BES) office contained three subprograms that originated in the ERDA Division of Physical Research. The three subprograms were: (1) Materials Sciences; (2) Molecular, Mathematical, and Geo-Sciences;

and (3) Nuclear Sciences. The Nuclear Sciences subprogram was dissolved in 1986 and its research programs were transferred to other offices within the DOE Office of Energy Research. The Office of Basic Energy Sciences underwent several structural modifications. and in 2003 was reorganized into its present form with three divisions: (1) the Materials Sciences and Engineering Division; (2) the Chemical Sciences, Geosciences, and Biosciences Division, and (3) the Scientific User Facilities Division. Today all external basic research grants are provided through the Office of Basic Energy Sciences.

In 2005, in response to a request from Congress, the National Academies conducted a study to "identify the most urgent challenges the U.S. faces in maintaining leadership in key areas of science and technology." The study also addressed issues of what specific actions policymakers could take to keep the USA at the forefront of science and technology. Their resulting report, *Rising Above the Gathering Storm: Energizing and Employing America for a Brighter Economic Future* [6] was issued in 2007. The report recommended that, following the Department of Defense success with the Defense Advanced Research Projects Agency (DARPA), a new office for energy research projects, the Advanced Research Projects Agency-Energy (ARPA-E), modeled after DARPA, be created within the DOE. Subsequently, ARPA-E was created in the America COMPETES Act in 2007, although no funding was provided. ARPA-E officially became functional in 2009 and in its first year provided $151 million for 37 research grants. ARPA-E does not provide support for basic research projects, and focuses its support towards technology-directed, applied research and development projects that address practical solutions to problems in energy creation, distribution, and use.

The National Institutes of Health (NIH) is one of the oldest government research organizations in the USA. It originated from a laboratory established in 1887 for the research of bacteria within the Marine Hospital Service (MHS), which had been established in 1798 for the care of sailors and merchant seamen. The world, primarily due to the work of medical scientists in Europe, was learning that bacteria and microscopic organisms were associated with the spread of infectious diseases. The laboratory established within the MHS was set up in Staten Island, New York, and moved to Washington, DC, in 1891 and was called the Hygienic Laboratory. The laboratory was officially reorganized in 1902 into four divisions, with the original bacteria and pathology research organized under the Division of Pathology and Bacteriology and the establishment of three new divisions, the Division of Chemistry, the Division of Pharmacology, and the Division of Zoology. The Hygienic Laboratory was officially renamed the National Institute of Health (NIH) by the Ransdell Act of 1930, which formalized the use of public funds for medical research. The National Cancer Institute (NCI) was established seven years later, with authorization to award grants to non-government scientists for research on cancer, as well as to fund fellowships for young researchers. The NCI was officially integrated into the NIH in 1944 and in succeeding years the NIH grew to 27 institutes and centers. In 1946 the NCI's grants program was extended to the entire NIH, with grants funding budgeted at a little over $4 million in 1947. NIH research grants funding increased rapidly over the years and was slightly over $10 billion in 1998. The NIH research budget was doubled by an act of Congress over the five-year period from 1998 to 2003, increasing from about $13 billion to about $26 billion. The amount of the NIH budget provided for grants research increased over this period from slightly more the $10 billion in 1998 to about $17 billion in 2003. It has decreased since then from the peak in 2003 to a little over $15 billion at the current time.

2.2 Industrial Support for Academic Research

When considering sources for academic research support an obvious question arises: Why doesn't industry provide the majority of research funds for university-based research? After all, the practical utilization of the research projects, when the projects are successful, is to provide products that are commercialized by industry. In this manner, the university-based research directly supports the economy and leads to new products, industrial innovation, and job creation. Industry, therefore, directly profits from the research. So why doesn't industry provide the majority of the research funds?

Historically, industry has, in fact, been both a major performer and supporter of research. Both small and large companies conduct research on a continuing basis, often with significant investment in both personnel and facilities. In fact, many small companies are started by means of research projects that result in new products. Large corporations and large companies conduct a very significant and sizeable body of research, often in company research centers. In the past, many of these industrial research centers performed a wide range of basic, applied, and advanced development research, with the end goal of improving their product line, as well as introducing new products to market. Many companies, both large and small, also work closely with academic scientists and engineers and provide a significant amount of financial support for academic research. However, the projects they support tend to be increasingly applied in nature, and focused upon product improvement and development. There has been a dramatic shift in research performed and supported by business and industry over the past three decades or so. Organizations such as IBM, GE, AT&T, etc., used to perform significant basic research, which in many cases was not necessarily directly related to their commercial operations. These large industrial organizations could take a long-range view of research, with anticipated downstream applications and effects. However, many of these organizations either closed down, or radically downsized their basic scientific and engineering research programs, and redirected the majority of their research efforts to more near-term and applied applications, with the focus upon near-term results. While many of the older large companies have shifted their emphasis, other and newer companies, such as Google, Microsoft, Tesla, SpaceX, etc., heavily invest in research, and many of these companies continue to support the research activities of academic partners and collaborators.

Research funded by industry has experienced a decline in focus upon basic science and engineering projects, and the trend towards more applied R&D is increasing. Many faculty members, particularly those in engineering disciplines, often work with colleagues in industry, sometimes in a very close relationship, on projects of mutual interest. The faculty member may work on industry-supported research in their university laboratory, or they may work as industrial consultants, in industry laboratories and facilities. These relationships are very beneficial to both parties and help the academic researcher get exposed to real-world problems, as well as state-of-the-art technology. This helps the academic researcher become more effective, which may lead to faculty members getting involved in commercializing their own research, and starting new companies. This has been an increasing trend over the past three decades and is encouraged by federal and state governments and by universities. An increasing number of universities have established incubators to nurture new start-up companies and to provide faculty with a variety of support mechanisms, which can include office and laboratory space, intellectual property assistance, and business support and training. These arrangements are developed by agreement between the university and their entrepreneurial faculty. These issues will be addressed in more detail later.

2.3 What We've Learned

Numerous sources of academic research funding have been established over the history of the country. While the US government is the primary source, and the government has established a diversity of agencies and offices that provide support for academic research, research funding is also available from local and state government organizations and agencies, private industry and businesses, and charitable and not-for-profit foundations. However, the research areas and topics, terms and conditions that guide and determine the available funding, and the amount of funding, will vary widely from source to source.

References

[1] https://en.wikipedia.org/wiki/Robert_H._Goddard
[2] *The Milwaukee Sentinel*, "Rocket Patent Suit Settled," Aug. 5, 1960

[3] David M. Hart, *Forged Consensus, Science, Technology, and Economic Policy in the United States, 1921–1953*, Princeton University Press, Princeton New Jersey, 1998

[4] K. T. Compton, "The Government's Responsibilities in Science," *Science*, pp. 347–355, April 1935

[5] Vannevar Bush, *Science: The Endless Frontier*, A report to the President on a program for postwar scientific research, July 1945. Re-published 1960 by the National Science Foundation in Washington, DC

[6] *Rising Above the Gathering Storm: Energizing and Employing America for a Brighter Economic Future*, Committee on Prospering in the Global Economy of the 21st Century: An Agenda for American Science and Technology; Committee on Science, Engineering, and Public Policy; National Academy of Sciences; National Academy of Engineering; Institute of Medicine, 2007

3 The Academic Recruitment Process: Position Announcement through Performance Reviews

In this chapter we'll discuss issues that a new faculty member will experience in the process of initiating his or her academic career. We'll start with a brief discussion of the academic search process mechanics and the various steps a university goes through when it seeks to fill a vacancy in its faculty or to add a new faculty member in order to expand or enhance its faculty areas of expertise. The seeds for academic success are often sown during the recruitment and employment negotiation portion of the search process for the new faculty member, and particularly during the negotiation of the start-up package. It's extremely important, in particular, for a new PhD graduate seeking an academic career to understand the recruitment process, because things move very rapidly once a position is offered, and well-considered decisions will need to be made in a timely manner. We'll also describe the expectations your institution will have regarding the performance of new faculty members and the academic performance review process. I'll indicate and describe the critical evaluation stages, and put these in order, in an effort to put into perspective the importance of making every attempt to quickly initiate a research program. What is commonly called "the start-up package" will be defined and described. This is particularly important since the items included in the start-up package, which is negotiated during the recruitment process, can have a very positive and significant effect upon research program initiation and development and, for this reason, the various items in the start-up package should be carefully considered. These items need to be defined and negotiated during the recruitment process and before a faculty position is formally accepted. It's very important to secure the necessary items and resources in order to get off to a quick start and be headed in the direction of success.

3.1 The Academic Recruitment Process

We'll start with a description of the academic faculty recruitment process and the various steps and stages involved. If you've already gone through this process and have received a job offer to become a faculty member in the department and university of your choice, you're to be congratulated. You've successfully navigated a very competitive, strenuous, and complex process. However, if you're just beginning the process and are starting to search for an appropriate faculty position that matches your interests and credentials, it should be helpful to you to understand the various stages and factors involved in the recruitment process. In either situation, you should find the information in this section useful.

First, you'll quickly learn that the recruitment process for a new faculty member is, in general, quite lengthy, as well as complex, involving numerous people in the department and at the institution performing the faculty search. Soon after submitting your application materials, you'll most likely receive an immediate response from the university indicating that your application has been received. However, after that you may wait weeks, or even months, before receiving any further response from the search committee or department chair. In some cases, you may never receive any further response, which of course, means that your application has been declined. It's unfortunate, but the failure to communicate with a faculty candidate sometimes occurs. This situation generally occurs with searches that are not well organized and miscommunication between the search committee and department management exists. Declined applications are sometimes lost in the shuffle, which is a very frustrating experience for job candidates. Nonetheless, this situation, although fairly rare, does occur, and a potential candidate should not be discouraged from applying for positions at other institutions and the failure to properly communicate is rarely related to the candidate's credentials. However, you should be aware that all searches take a significant time to complete, and the search process generally will occupy the major portion of an entire year before it is successfully completed.

To conduct a new faculty search, a vacancy must exist, or be anticipated in the near future, within the department, and this could result from the retirement or departure of an existing faculty member, or the creation of a new position due to expansion of the department faculty. The number of faculty positions is determined and specified by the university, usually under the control of the provost, and the faculty positions are managed by the dean of each college within the university. Faculty positions are assigned to each department, and the number of faculty positions allotted to a department is based on a number of factors including undergraduate and graduate student enrollment numbers, subject area, and specialized needs. Each department will generally define its own faculty needs and the department head or chair will negotiate with the dean and make requests for additional or new faculty members in response to an anticipated vacancy. In some departments the faculty positions are closely identified with specific subject areas and maintained within the department and, when a vacancy occurs, the department will generally be authorized to conduct a search for a replacement faculty. However, in other colleges the faculty member positions are more tightly managed by the college and the dean may decide to not fill a vacancy that occurs in a particular department, but rather move the vacancy to another department. In this manner the dean has the ability to manage the college staffing and build the size of certain departments while decreasing the size of others. This generally occurs when a specific discipline is experiencing growth, while another discipline is in a stagnant or declining environment. Of course, the dean can also elect to not fill a vacancy, which will often occur during periods of financial stress and budget uncertainty. This flexibility is very important for a dean to be able to effectively and efficiently manage the college. In general, the college dean controls the faculty positions in each department and has the responsibility for authorizing new faculty searches.

Once a faculty vacancy has occurred and a new faculty member search authorized, the new position opportunity will be announced and published. Typical publications that routinely publish new faculty openings include *The Chronicle of Higher Education, AcademicKeys,* and most of the technical journals that serve professional societies. There are also numerous websites devoted to academic job openings that regularly publish current faculty and academic professional openings, and some popular websites include:

- http://jobs.sciencecareers.org
- https://academicjobsonline.org/ajo
- http://www.nature.com/naturejobs/science/
- http://www.academiccareersandjobs.info
- http://www.academic360.com/default.cfm
- http://www.academickeys.com

Also, universities will normally publish their new faculty searches, indicating their open faculty positions and related information, as well as application requirements on their websites, so it's always a good idea to search the various department websites, especially if you're interested in a particular university. In these announcements and website postings they'll indicate the rank of the open position, the discipline, areas of specific expertise, the expected time period of the search and the closing date for candidate application, and other associated factors of interest. They will also often include background or general information regarding the department, college, and university, and sometimes information related to the local area. Other mechanisms for advertising open faculty positions are also employed, and it is common for departments to circulate announcements of their faculty vacancies to their peer institutions. Some announcements will indicate specific expertise topics of interest, while other searches will be directed towards identification of the best and most highly qualified candidate, no matter the area of special expertise. In the latter case, they will generally list the subject areas that are offered within the department.

The process of searching for a new faculty member is fairly consistent throughout academia, and effectively follows a similar and uniform procedure. The basic procedure, although it could vary somewhat depending upon the specific institution, is complex and lengthy, and will consist of the following steps: (1) formation of a new faculty search committee to execute the search; (2) announcement and advertisement of the vacancy; (3) collection of candidate resumes; (4) evaluation of the resumes by the search committee; (5) down selection by the search committee members of all candidates under consideration to a limited number of highly qualified candidates; (6) solicitation of external letters of recommendation for the highly ranked candidates; (7) identification of the top three to five candidates (sometimes more); (8) invitation of some or all of the top-rated candidates to a campus visit, during which the candidate will be asked to make a presentation on their current research and future plans, and interviews with the search committee members, department faculty, and department leadership; (9) collection of candidate evaluation comments from department faculty by the search committee members, and determination of a list of the candidates deemed most suitable for the advertised position; (10) submission of a ranked (sometimes unranked) list of recommended and most highly qualified candidates to the department head or chair; (11) consideration by the department head or chair of the list of top candidates and determination of the top candidate to be offered the faculty position; (12) possible invitation (this step does not always occur) to the selected candidate for a second on-campus visit and informal discussions, where the selected candidate will meet primarily with department leadership, possibly faculty members with similar research interests to the selected candidate, and possibly the college dean; (13) issue of an offer letter for employment by the department head or chair to the selected candidate (most often the department head or chair will make the final selection decision); and (14) negotiation of the faculty position offer and the associated conditions with the department head or chair and the selected candidate.

The final candidate selection, depending upon the institution and discipline, may be the responsibility of a faculty committee, the department head or chair, or the college dean. However, in most colleges and universities, particularly the large institutions heavily engaged in research, the final decision for a new faculty hire is a highly valued responsibility of the department head or chair since this is the primary mechanism by which they have the ability to build their faculty and focus their program into developing and emerging technical areas, or to gain expertise in an area they feel is deficient in their department. The entire search process, as indicated, is quite lengthy with numerous steps, and the search can easily extend over many months and, most likely, will occupy the majority of an academic year. The search process will generally begin in the early fall, with the final faculty position offer being extended to the selected candidate late in the spring semester, with the intention of having the new faculty member join the department either in the summer months, and in time to begin their new duties in the fall semester (i.e., August or September). Other start time arrangements can be negotiated with the department head or chair, and it is possible to begin employment in the winter or spring terms, generally in January. However, employment will usually be synchronized with the academic year, with the vast majority of new faculty starting their employment either in the fall or spring semesters.

The search process for a new faculty member is also, in general, very competitive. In the current environment it is not uncommon for an announced faculty vacancy to receive anywhere from a relatively small number of candidate applications, in the range of 20–30, to a very large number, in the range of 200–500, or more, depending upon the specific college or university conducting the search. New faculty search committees will receive applications not only from candidates with residence in the United States, but also from candidates located worldwide. In general, and at most US institutions, there is no formal restriction related to recruitment of faculty from international institutions, and lack of US citizenship is not a limitation for the candidate. However, candidates that do not live in the USA may be at a disadvantage, particularly since travel for campus visits during the recruitment process is paid for by the institution, and international travel costs, visas, etc., are an additional consideration for the search committee, which may have a restricted or limited budget for the recruitment process. If a non-US citizen is hired, the university will go through the process of applying for the necessary documents required for that person to legally be employed in the United States. Also, US university graduate schools enroll a large number of international students, and many of these new PhD recipients will apply for faculty positions within the United States.

Many of these applications are successful, and this is the main mechanism by which foreign nationals become faculty members at US institutions.

The search committees, understandably, have a very difficult time in evaluating the large number of resumes that are submitted. This is a very critical stage in the search process and the search committee members will carefully evaluate all information submitted so that they can identify and determine the final list of highly qualified and top-rated candidates that will be invited to the university for a campus visit and job interview. The time required for the detailed evaluation is one of the major reasons for the lengthy period of time it sometimes takes for a candidate to receive information regarding the status of their application. A significant delay in receiving any communication after applying for a position is not necessarily an indication that your application is not receiving a favorable review and evaluation. In fact, the delay could indicate the exact opposite. Notifications of the applications that are declined are often sent to the candidate early in the process. Therefore, in order to receive a favorable review of your application, it is necessary to carefully prepare the material you submit. It is extremely important and necessary to prepare an effective and well-written resume.

In the application materials you submit to the search committee you often will be requested to include: (1) a cover letter describing your interest in the position; (2) a copy of your resume or CV, indicating your education and training, your employment history, your professional activities, associations, and memberships, and a list of your publications, conference and workshop presentations, and other information related to your professional performance; (3) a list of professional references, including complete contact information, who have agreed to provide you with a letter of recommendation addressing your status, reputation, and performance in the professional community; (4) a statement of your research and education program interests and development plans; and (5) copies of one to three of your published journal or conference papers. This list is representative of items that are requested from applicants, although the actual items requested may vary from institution to institution and from discipline to discipline. It is important to include all items requested by the search committee, and to eliminate, or certainly minimize, extraneous information not specifically requested. Failure to include requested information is often the grounds for declination of the candidate, particularly when there are many candidates under consideration. Concise and effective writing is important and all requested information should be presented in a clear, well-written, and brief manner. Search committee members, particularly when they have a large number of applications to review, will not necessarily spend time or give a thorough review to poorly written applications or support material provided by the candidate.

3.2 During the Interview and After Being Offered a Position

After the search committee has determined a short list of finalist candidates, a small number of them, generally in the range of three to five, will be contacted and invited for an on-campus interview. The institution will provide the travel expenses for the visit, and they will contact you regarding appropriate travel arrangements, as well as provide you with an itinerary and agenda for your visit. You will also, generally, be provided with brochures and information concerning the university and other information related to living in the area where the institution is located. Once you have been selected for an on-campus visit and interview, you will have the opportunity to discuss your possible addition to the faculty with various members of the search committee, the department faculty, the department chair or department head, and, most likely, the dean of the college and/or members of the dean's staff. You will also be asked to give a presentation to the department faculty, including the search committee members, of your current research activities, and your plans to continue your research. The research presentation, which consists of your presentation followed by a question and answer period, is generally limited to an hour and should be carefully planned since it is very important.

The research presentation is the main mechanism for you to make your case to the department faculty for how you will fit into the department regarding research interests and what new dimension and/or direction you will add to the faculty. The search committee and department faculty are generally looking for someone with specific research interests and you want to make the case to them that you are the appropriate candidate to satisfy their needs. For this reason, it's best to inquire of the person contacting you to invite you for the campus visit if there are specific topics you should address in your presentation. If the person who contacts you is not the department head or chair, you should contact the department head or chair to clarify their expectations for your visit and presentation. It's important to keep in mind that the department head or chair is, most often, the person with the authorization to make a faculty position offer. Also, you should research the department and identify specific research areas of expertise and, particularly, research groups and faculty members with whom you have mutual interests and with whom you would expect to collaborate. This is easily accomplished by means of a review of the appropriate websites, where research faculty and research groups typically indicate their areas of expertise and recent or most important professional contributions. Also, the websites will generally indicate the most significant publications for the various faculty members, research groups, and research laboratories. Keep in mind that many faculty members attending your presentation will form their opinion of your suitability for the position by your performance during the presentation, and they will rate your performance for both research and teaching by their evaluation of how effectively you present your research and your overall ability to communicate. These faculty members will be asked to provide their evaluations and comments to the search committee for use in the search committee's candidate evaluation and ranking process.

During your discussions with individual faculty members and with the department and college leadership you will have the opportunity to ask questions about the position, clarify your duties and responsibilities, and to inquire concerning department and college resources that you will be provided. One of the most important issues associated with accepting a faculty position, particularly for new and young people, but also applies to all faculty members being recruited to faculty positions whether they are new PhDs or experienced and mature faculty members, relates to start-up packages. The start-up package refers to resources that are promised and provided to new faculty members when they are recruited and before they are hired. The start-up package is a temporary allocation of resources that will be time-limited, and is generally available over a period of time commonly ranging from one to three years and possibly to five years, depending upon the institution and specific needs identified in the negotiation. After this time the start-up package will terminate. The details of the start-up package should be discussed during the interview process, particularly if a second on-campus interview is offered, and the details should be agreed upon and finalized so that these items are included in the offer letter. Once the details of the start-up package are agreed upon and finalized, it is very difficult to change and, therefore, careful attention is necessary to make sure the appropriate and proper items have been included.

The resources included in the start-up package are used in the new faculty member's efforts to initiate and/or establish a research and education program, and are, therefore, very important to the new faculty member since the start-up package items will directly affect the faculty member's initial professional development and career progress. The start-up package may consist of a lump sum of money provided on a one-time basis, or it may consist of a list of specific items. The contents of the start-up package will vary from individual to individual, and also from institution to institution, and the exact items that are included in the start-up package are generally negotiated between the new faculty member and the chair or head of the department in which the faculty member will work, although often the college dean may also be involved. The amount of funding available for start-up packages for new faculty members will vary from university to university, but virtually all colleges and universities now offer some sort of start-up package to new faculty members. Also, the flexibility and degree of negotiation regarding what is to be included within the start-up package will also vary from university to university and in some cases may be limited, with little flexibility, while other situations may offer a considerable range of options. The financial resources available to academic institutions for start-up packages are generally derived from funds obtained from endowments, university fund-raising, overhead returns from research grants, and other activities not related to the institution's normal operating budget. Research-intensive institutions tend to offer start-up packages with relatively high dollar amounts and with a corresponding wide range of options, while institutions that are more oriented towards teaching activities may offer start-up packages with fewer options and lower overall dollar amounts. For example, a top-rated and large research-intensive university may offer a start-up package in excess of a million dollars, while a smaller, regional university may only offer a start-up package on the order of $25 000 to $50 000, or less. Higher monetary values for a start-up package are often related to experimentally intense research that requires specialized equipment or instrumentation not readily available in the department.

However, no matter the institution, when negotiating a faculty position, careful attention to the start-up package is important from a variety of perspectives, but primarily from the perspective of building a research program. The start-up package can provide you with necessary support in the form of materials and supplies, or research equipment required for you to pursue your research, which is particularly important if you require a specialized equipment item that is not already available at the university. The start-up package will also typically include support for student researchers, time release from academic duties (sometimes called "academic year funding"), such as teaching and committee work, and, very importantly, travel support to enable you to visit research funding agencies and/or professional conferences and meetings. This start-up support will, most likely, prove instrumental in your early attempts to perform initial and preliminary research and help you obtain some results that you can use in proposals for research grant support. Therefore, the resources included in the start-up package are important and will have a significant effect upon your efforts to obtain research grant funding. Exactly what you should ask for in your start-up package should be carefully considered.

The resources included in the start-up package will also affect your status in your new academic position and your performance reviews. You will learn that the time you are provided to establish your academic career is limited and on a formal and defined time schedule, and that your critical performance evaluations loom on the calendar. The moment you accept your academic faculty member job offer, the clock starts ticking and your first critical performance evaluation is rapidly approaching, and you need to quickly demonstrate evidence that you are developing and progressing in the correct direction. You need to hit the ground running and to get research results as soon as possible. This is particularly important since initial research results are highly valued by program managers, program directors, and proposal reviewers in grant funding agencies, and initial research results are often a key factor in receiving favorable reviews of your proposal and obtaining grant funding, and, therefore, this also increases the pressure and need to get your research established and for your research to produce valuable results at the earliest possible time. This issue is described in more detail in Chapter 6. Your start-up package can have very important and positive effects upon this process and the initiation of your research in your new institution. Although the resources included in the start-up package are very important to you and will affect your performance progress, exactly what is best to negotiate in the start-up package can vary, depending upon your career path and interests, and the contents of your start-up package should be carefully considered. Since the amount of funding available for a start-up package may be limited at your university, it is best to prioritize what you wish included in your start-up package to facilitate your negotiation. As we'll see, from the perspective of building a research program, some items are far more important than others, and it's best to identify what will be most important to you. You may not be able to successfully negotiate everything that you would like included.

3.3 Performance Reviews

Before discussing the offer letter and start-up package in detail, it's useful to review the guidelines and process for performance reviews that you will experience, and the general procedure for promotion associated with an academic faculty position. The performance of all faculty members is evaluated on an annual basis, but new and non-tenured faculty members at the Assistant Professor rank face critical performance reviews in the third (sometimes the second) and sixth year of their employment. The mid-term evaluation at the end of the third year is to evaluate the faculty member's performance for re-appointment at the Assistant Professor rank, and the evaluation at the end of the sixth year is both intense and critical since it is for consideration of granting permanent tenure and, usually, promotion to Associate Professor. All faculty members must either be granted permanent tenure within a seven-year time period, or be released from their employment. Therefore, the sixth-year performance review is the single most important, and detailed, evaluation a faculty member will face in his or her career. After the third-year and/or sixth-year evaluations, the university has the option to terminate the faculty member's employment in the event that their progress is deemed unsatisfactory. In this case the faculty member will be given a fourth year and/or a seventh year, depending upon the review, so that the faculty member has time to search for another position. After a negative review in either of these two evaluations, the faculty member will no longer be considered on the tenure track, and any continued employment by the institution will be in a non-tenure track position such as Lecturer or Researcher. However, continued employment is not guaranteed, and the majority of faculty members denied tenure leave the university.

The performance reviews you will experience are mandatory and formally defined by your college or university. The performance evaluation guidelines establish clear expectations concerning how you and your performance progress will be evaluated, and by whom. In your discussions with your department head and/or faculty mentor, these expectations will be addressed in detail, and both you and your department head will agree on reasonable goals for you to pursue. At some institutions your interactions with your department head or faculty mentor may be informal. However, it's becoming more common for these interactions to be formal, with specific goals and expectations to be agreed between you and your supervisor, and written in a formal document and indicated as a Statement of Mutual Expectations, or some other document with an equivalent title, which will be signed and dated by you and your department head and included in your personal file. Your subsequent performance reviews may make use of this document when your supervisor evaluates your performance progress.

The start-up package you negotiate will have a significant effect upon how you are able to function in the first few critical years of your academic employment, and the resources provided will assist you in transitioning to an academic career. The university, college, and department administration, along with the department faculty, will make every effort to assist in this transition and to provide guidance to you. Formal procedures for progressing in an academic faculty career, which are essentially the same at all colleges and universities, have been established, and during your first few years, and up to a maximum of seven years, you will be employed in a probationary position and you will be subject to annual performance reviews. Before these annual performance reviews the faculty member, whether tenured or untenured, is generally asked to complete a document that lists all the activities in which they've been involved in the previous year. The list is comprehensive and can significantly vary, depending upon the faculty member and their assigned and selected activities. These documents are generally termed the Faculty Annual Report, or some similar title. These reports become a part of one's permanent record in the department files. This document generally provides information to your supervisor and serves as the basis for the annual performance review and possible salary adjustments.

As mentioned, for new, untenured faculty members, the most critical review occurs when they are reviewed and evaluated for promotion to Associate Professor and the award of permanent tenure. This review can occur at essentially any time once they have commenced their employment, but must occur no later than the sixth year of their employment. If a decision to not award permanent tenure is made, the faculty member is still employed for a seventh year, which provides some time to find another employment opportunity. If a faculty member is making good progress in their career, an early critical tenure review before the sixth year is possible. In most academic institutions, with the possible exception of departments with very large numbers of faculty members, the performance of all junior and pre-tenured faculty members is discussed by all the tenured department faculty, and all will have a vote, generally indicating whether the untenured faculty member is "meeting expectations," or "not meeting expectations" in his or her performance. Large departments will often have a special committee of tenured faculty members established to evaluate junior faculty members for promotion and the award of tenure. In either case, the junior faculty members that are making good progress are identified, along with pre-tenured faculty members that require some additional assistance. This process, along with the faculty member's annual performance review with the department head or chair, can result in a recommendation that the faculty member be considered for early tenure, and it's possible that the faculty member will undergo a full permanent tenure review before the sixth year. Also, the faculty member can independently request that they be considered for tenure and promotion at essentially any time during their probationary period. However, it's always best to discuss the timing for this critical review with your supervisor and/or department head or chair, and an early tenure review is not always a good idea since the review may not prove successful. Additional time may be required in order to provide the opportunity to add to your record and work towards correcting any perceived deficiency. Again, the best advice is to discuss your situation and performance with your supervisor. However, with consideration of these caveats, it's not at all uncommon for the tenure review to occur before the sixth year.

Once a pre-tenured faculty member has been notified by their department head or chair that he or she has been recommended for the permanent tenure and possible promotion review, the faculty member will be requested to prepare a formal tenure evaluation dossier. In this dossier they will list the complete spectrum of their activities since accepting his or her academic position. The dossier contents are very similar to the faculty annual review report, with the exception that the activities will include all activities since the acceptance of the faculty job offer, and the initiation of the faculty member's academic career. Also, the dossier will generally include certain corroborating information such as copies of selected journal articles you've published, or other evidence of academic achievement. The dossier will also include letters of recommendation solicited from references identified as experts in your field. The experts could be faculty members at other universities, or scientific experts from government or industrial organizations. You will be able to suggest names of experts you feel appropriate to serve as a reference for you, and the list may include people you know personally, or experts you know by reputation. The list of possible experts to serve as references will be determined by members of the tenure evaluation committee or your department head or chair, and often will include some people that you have not suggested and do not know personally, but are identified as experts in your field by the department faculty members serving on the promotion and tenure committee, and/or the department head or chair.

The people selected to provide letters of recommendation will be chosen on the basis of their professional reputation and their ability and willingness to comment on your work, generally by a review of your published papers and other information included in your resume. For example, in addition to your published research work, your record of obtaining research grant funding, and your participation in professional activities are often topics addressed by the outside experts. The exact list of people that will provide the letters of recommendation regarding your possible promotion and award of tenure will be selected by the chair of the promotion and tenure committee responsible for evaluating your credentials, or the chair or head of your department. They will contact the references and solicit the letters of recommendation, and you will not be directly involved in the process. Generally, the references will be provided with an assurance that their identity will be protected to the extent permitted by university policy or state law. Therefore, you will not, most likely, know their identity. The people providing the letters of recommendation will be provided copies of your resume or CV and copies of your selected publications. They will be asked to comment on information included in your CV, and the quality of the content in your published papers, and then requested to comment on their evaluation of your progress and status relative to other faculty members they know, generally at their own institution, and at a similar point in their career. The letters will be included in your dossier and will be used in the promotion and award of tenure evaluation process. Since the evaluation by outside professional experts is an important component of the performance evaluation and tenure process, you should make every effort to develop professional relationships with as many of your colleagues as possible.

After the award of permanent tenure, a faculty member is still subject to annual performance reviews, but these are primarily for the purpose of reviewing the previous year's activities, and to provide a forum for a personal discussion with the department head or chair, and gather information to make decisions regarding salary adjustments, etc.

3.4 Permanent Tenure

As indicated, permanent tenure is a major goal to work towards in an academic career, but what is permanent academic tenure and why is it so important? Employment in industry or other private institutions does not involve tenure. What is different about academic faculty employment, and why does tenure exist? The answer to this question is based in history and the evolution of academic institutions as a venue for the free discussion of ideas and the free expression of thought, without undue restrictions for political or other reasons. Although academic tenure has a long history, the modern system essentially began in 1940 when the American Association of University Professors (AAUP) published a "Statement of Principles," which outlined two major goals, indicating conditions that should be established and under which tenured faculty members should function. These two goals were that the faculty member should have (1) "freedom of teaching and research," and that they should also have (2) "a sufficient degree of economic security to make the profession attractive." In the modern period tenure is generally considered under the first principle, and the second principle is largely forgotten. The award of permanent tenure is highly valued, and is often used as a recruitment factor, particularly when someone with outstanding credentials and employed in government or industrial organizations is being recruited to an academic faculty or administration position.

The AAUP "Statement of Principles" also recommended the current seven-year probationary period, and suggested that a tenured professor could not be dismissed without reason or the chance for a self-defense opportunity where all factors associated with the dismissal be heard. Tenure provides for academic freedom and the right of faculty members to pursue their research and teaching interests. Therefore, permanent academic tenure, although not absolute, provides a certain degree of job security. Tenure is associated with the department where the faculty member is employed. Once tenure is awarded, a faculty member can be dismissed and his or her employment terminated only for adequate cause or other specified circumstances, and only after a hearing before a faculty committee of peers, and following due process procedures where all facts of the possible termination are thoroughly considered. During the hearing the faculty member may be assisted by a faculty advocate or legal counsel. The due process procedure generally is concluded with a decision by the university provost and president or chancellor.

Tenure, of course, does not give protection for criminal convictions or other illegal activities, or other disciplinary actions, which can result in dismissal for cause. Also tenure does not protect a faculty member from job termination due to the closure of the faculty member's department or other financial disruptions the department, college, or university may experience. However, tenure does guarantee due process and ensures that there is an appropriate forum where all factors will be considered. Tenure is highly valued by both faculty and academic administrators and is awarded only after the new faculty member has served a probationary period and after a thorough and complete multi-level performance evaluation process. Once tenure is granted, it usually stays with the faculty member throughout his or her career, even if there is a change of employment to a different university. That is, if an Associate Professor or Full Professor with tenure is recruited to a new university, he or she is generally offered a similar and tenured position in the new university. If a faculty member resigns from the university to accept a position in an industrial, business, or government organization and then later wishes to return to an academic position, tenure is not guaranteed and would need to be negotiated with the new academic institution. Tenure in this case may or may not be offered. The circumstances and procedure associated with the award of permanent tenure, along with the faculty member's duties, rights, and related processes, are clearly described and defined in the faculty handbook, or other documentation, that is provided to a new faculty member when they are hired. The handbook and other policies associated with an academic faculty position will generally be listed on the university's website, under Human Resources or the Provost's Office organizations.

3.5 The Probationary Academic Review Process

In this section we'll discuss the academic review process, and demonstrate the importance of negotiating a start-up package that will enhance your ability to get a quick and effective initiation of your research. First, if you've already been offered a faculty position you are to be congratulated since you've made the most important first step on the long path of your academic career. If you've recently graduated with your PhD, or have one or two years of post-doctoral or industrial experience, you have most likely been recruited for a position at the Assistant Professor rank. If you're already at the Assistant Professor rank in another university, or if you have several years of industrial experience, you may be recruited at the rank of Associate Professor, possibly with tenure, although most likely without permanent tenure. An offer of permanent tenure is generally made only if you have already established a successful and outstanding academic and/or research record, either at another university or as a mature industrial or government scientist or administrator. If your research record is truly outstanding, you may be recruited at the Full Professor rank, which would almost always include an offer of permanent tenure. However, these situations are relatively rare, and restricted to mature scientists, engineers, and administrators, generally with outstanding and extensive records of research, service, and publication.

In order to explain the importance of the start-up package, we'll assume you're a recent PhD graduate, you've successfully navigated the recruitment process, and you have been rewarded with an offer for a tenure-track faculty position at the rank of Assistant Professor at a research-intensive university. Your acceptance of this position is a necessary first step in your future career. As a recent PhD graduate, you are now beginning a process that includes a pre-tenure, probationary period, as previously indicated, where your academic and research performance will be carefully evaluated through a very comprehensive and detailed process that involves the tenured faculty in your department, your department head, and, at the critical times, the dean of your college, representatives of other departments within your college, possibly representatives from other colleges within the university, and then administrators on the college and university level, such as the provost. The formal evaluations occur on an annual basis, as previously discussed, with critical evaluations occurring in the third and sixth year of your employment. The annual evaluation is generally performed by your department head or department chair, and usually based upon your completion of a Faculty Annual Report (FAR), or some equivalent document. In this document you will list virtually everything you've done professionally the previous year. Generally, these activities are grouped into three major categories of (1) teaching, (2) research, and (3) service. The exact ranges of activities included under these three categories are loosely defined, and can vary and will periodically change, and will be interpreted in the context of your job description and expectations.

The types and range of activities generally considered under the "teaching" category include activities such as courses you have taught the previous year, new courses you've developed and introduced, existing courses you've significantly modified, textbooks you've written, your involvement with online and distance education courses and activities, students you've mentored and advised, and other efforts related to participation and support of educational and teaching activities. You may have been involved with teaching a special course to industrial students, either at your university or at their industrial location. You may have been involved in tutoring activities for student organizations, or volunteered to tutor students having difficulty in their courses. In general, any activity that is involved in your educational and teaching efforts should be listed. Also, this is just a partial list of common activities generally considered under the "teaching" category and not all activities may be applicable or appropriate for your position.

The "research" category includes an accounting of your research interests and pursuits, and includes a listing of research projects to which you've contributed your time and effort, whether the research projects are unfunded, externally funded, or supported with internal funds. You may be asked to indicate your level of involvement as a principal investigator (PI) or co-principal investigator, or as a contributor on externally funded research programs. Often you'll be asked to indicate the percentage of your time committed to the project, and you may be asked to list the total dollar value of each project, along with a list of your "personal share" of the grant funds for each research grant award. This often occurs when there are multiple faculty members listed as PIs or contributors to a research grant project and the grant funds are allocated among the various participants. In their junior faculty performance evaluations, the promotion and tenure committee members and department faculty often value the principal investigator of a research grant as more significant than a participating faculty researcher, even though the participating faculty member may be, in fact, making very significant and important contributions. Therefore, it's important to focus effort upon building your own research program, as well as collaborating with other faculty members, when appropriate and mutually beneficial.

You'll also be asked to list research proposals you've prepared and submitted to fund granting agencies, research grants and the amount of funding you've successfully obtained, a listing of research papers you have in preparation, and those you've had published in peer-reviewed professional journals, a listing of papers you've published in non-peer-reviewed magazines, a listing of papers you've presented at professional meetings, conferences, and workshops, a listing of your efforts to build your research support facilities, including laboratory and research equipment you may have acquired, and other factors and activities associated with the performance of research and related efforts to build research infrastructure. You may also be asked to indicate if you are participating in research centers or laboratories that may exist, either at your university, or at a collaborative institution. The research category also includes your recruitment of graduate student researchers, and your mentoring and advising activities. You'll be asked to list the graduate students with whom you are associated and your level of involvement with them, such as Chair or Member of their MS or PhD advisory committee. You'll also list students you directly advise and mentor, and any post-doctoral researchers or visiting researchers you have hosted or supported.

The "service" category is normally very general and vague, and can include a very wide range of activities. "Service" activities can range from local community and civic activities to involvement in professional society organizations. Your "service" duties include your involvement on department, college, and university committees. The management and administration of academic organizations are heavily dependent upon a large number of committees, made up of faculty and sometimes staff members, who are expected, as part of their official duties, to participate in the committee activities. Committees are established for essentially all the critical functions that are associated with the operation of an academic department. Typically, committees exist for undergraduate affairs, graduate affairs, faculty recruitment, student groups and organizations, community liaison and outreach, academic program reviews, technical area groups within the department, junior faculty tenure evaluation, library and academic resources, teaching laboratories and equipment, student awards, faculty and staff awards, and other subjects and topics that require organization and management.

Committees also exist on the college and university level, and the entire committee structure and network can be quite extensive and comprehensive. Many universities have a special committee, with a name such as the Committee on Committees, or something similar, in order to effectively manage the extensive committee network and process. You'll be surprised and amazed by the number and range of responsibility of the various committees. The majority of committees are permanent and meet on a continuing basis, while some committees are *ad hoc* and organized to address a specific and particular one-time or short-term issue that needs to be resolved. The modern university could not function without the input and actions provided by faculty and staff through the committee system, and all faculty members, no matter their rank, are expected to spend some of their time serving on some of these committees. Committee participation is widely considered a general requirement and responsibility of faculty service, and participation on committees will be listed as a "service" contribution in your faculty annual report. Participation on a committee may be by appointment by your department head or chair, or by faculty election. Often, volunteers for certain committee assignments are solicited, and faculty members have the option to select committee service in an area that they feel appropriate for their interests.

"Service" activities can also include working with students in local schools, and volunteering to work with students and student organizations in elementary schools, middle schools, high schools, etc., is becoming more and more common for university faculty members. This activity is generally termed "outreach" and is very highly valued by certain grant funding agencies, such as the National Science Foundation. Outreach activities should be listed not only under the "service" category on your Faculty Annual Report, but also on NSF proposals under the "Broader Impacts" area of the proposal. In fact, these activities are so highly valued by NSF that they offer grant-funding opportunities to support and encourage participation of university faculty members in these efforts. This subject will be covered in more detail in Chapter 7. Other very important service activities include involvement and participation in professional society organizations and their various functions and events. Generally, this starts as a "volunteer" activity to assist in technical reviewing activities associated with a professional journal or conference. The professional societies require a large number of volunteers to staff and operate the wide diversity of activities they offer and support. Professional societies and organizations consist of people employed and engaged in like-minded professional pursuits, and these organizations organize and offer many events and services to support their members. Service in one of these professional organizations is a highly valued service activity, and your department and college will encourage participation and, often, provide some support for your volunteer participation in the professional organization.

The range of possible volunteer functions in professional societies and organizations is wide and diverse, and ranges from reviewing activities, to conference and workshop organization and management, to editorial service with a professional journal, to management and administration of the professional society. In fact, many faculty members devote a significant fraction of their time and effort to professional society service and participation, generally, by serving in numerous and diverse positions over a sustained period of time. Service in professional societies, and the various functions and venues they offer and support, is highly recommended and the service can provide excellent opportunities to meet both colleagues with similar scientific and technical interests from other institutions, as well as program managers and program directors associated with various grant funding agencies. Serving on an organizing committee or session chair for a conference or workshop can provide the opportunity to invite speakers from funding agencies; this, in turn, can provide not only the opportunity to learn of initiatives and funding priorities in their agencies, but also to get the chance to personally meet them and exchange ideas and thoughts. As emphasized throughout this volume, personal contact with funding agency program managers and program directors is of paramount importance, and you should pursue every opportunity available to meet and discuss research thoughts and ideas with them and to build your personal network of professional colleagues.

3.6 The Start-Up Package

At this point we'll return to the issue of the start-up package and provide additional details. During your interview visit and discussions with your new department head, and possibly the dean, you probably were questioned regarding special needs and expectations you may have regarding initiation of your research. This question is generally asked of researchers primarily and heavily engaged in experimental research that may require specialized laboratory equipment and instrumentation. The department head or dean will want to understand what sort of support they are going to need to provide. You may also have been questioned regarding your salary expectations. These questions are fairly standard for academic administrators and managers to ask new faculty members in the recruitment process. Sometimes new faculty members do not feel comfortable providing honest answers since they fear that an unreasonable answer may hurt or diminish their opportunity to actually receive an offer of employment. However, you should be prepared to have an open discussion and to offer honest answers since it's best to initiate your relationship with your new institution and your new colleagues with a clear understanding of what sort of resources you expect to be provided, and let them know how you propose to proceed to build your program. Discussion of the resources that will be provided to you involves the negotiation of your start-up package, and it's best to consider your options and requirements before you actually are made a job offer and negotiate your new position. There is always some flexibility in the resources and conditions associated with a new faculty position during the recruitment process, but it's generally much more difficult, and sometimes impossible, to obtain additional resources after a faculty position offer has been made and accepted. The flexibility for modifications to an offer in the post-acceptance period is severely limited. Therefore, you should carefully consider your needs and requirements before entering into the negotiation period, and have a clear priority of the resources you wish to have included in your start-up package.

In this section we'll discuss items you should consider including in your negotiation regarding your start-up package. Items we'll discuss include: your salary, internal university-provided items such as travel funds and a discretionary account, student tuition and stipend support, access to laboratory and computing facilities, funds to purchase special research equipment and instrumentation, summer salary support for yourself, and, possibly, temporary release time from teaching duties, and temporary reduction or delay in committee service. You may also wish to clarify your rights to perform outside consulting for pay and the relevant university policies associated with external consulting activities. What you negotiate in your start-up package should be provided to you in writing, and will generally be included in your employment offer letter. Your offer letter will include your salary and the university policy regarding obtaining salary for the summer months, the details of the start-up package, and a statement regarding the university's policy regarding outside consulting privileges. The offer letter will also state the department's expectations regarding your teaching and research efforts, and indicate any temporary reduction in your teaching or committee service expectations. Other special arrangements that you negotiate will also be listed.

The various items appropriate for inclusion in a start-up package are discussed in more detail below.

3.6.1 Salary

The very first item you will likely consider in your job offer negotiation is salary. You may think that this is the most important issue that needs to be settled, but although salary is obviously important to you, it is actually down the priority list for your long-term success. University department heads and chairs, and college deans, are well aware of current salary ranges for faculty members in the various discipline fields, and at their own and peer institutions, both on a regional as well as national level. They obviously are well aware of the salaries for current employees in their organizations since it is their duty at most institutions to set salaries and determine annual salary increases that may be awarded. At some institutions a peer committee determines faculty annual salary raises, but this situation is not the norm. Also, there are numerous surveys, published by a variety of professional organizations, of salaries of faculty members at various ranks in the various types of academic institutions, and these surveys make current salary data on both regional and national bases readily available. Also, department heads and chairs often participate in organizations of their peers where issues such as faculty salary, factors associated with tenure and promotion, issues associated with teaching and research, issues associated with laboratory facilities and maintenance, factors involved with faculty and program development, and other development and management issues associated with an academic and research program, are routinely discussed. Also, department heads and chairs and their college deans are involved in recruiting new faculty on a fairly regular basis, and in some cases they conduct searches for multiple new faculty members on a yearly basis. For example, academic departments with relatively large numbers of faculty members often have open searches for multiple faculty members in selected subject focus areas in any given year. College deans, in particular, will likely be involved in multiple new faculty member searches in the various departments within their college, and they will be well aware of current salary ranges for new faculty members at all ranks. Therefore, the salary offer you receive will have already been considered by your department head or department chair, and most likely will have already been discussed and approved by the dean of the college. The salary offer will, most likely, be a competitive salary and consistent with what new faculty members in similar positions to your situation and rank are offered.

Also, keep in mind that you have gone through a very competitive recruitment process, and if you are being offered a faculty position, the emphasis of the university faculty and administration is to get you to accept the offer and to join their faculty. In this regard the department head or chair is actually serving as your advocate and is working to get you the best offer possible. They have already expended considerable time and effort in your recruitment, and the last thing they want to do is insult you with a low salary offer, with the likely result that you do not accept the faculty position offer. In the rare case where you are being recruited by several institutions, you may have already received an offer to join the faculty at one institution, but if this institution is not your top choice, you can use the salary offer you have received as leverage in your negotiations with your desired institution. In this case you should share your job offer information with the department head or chair at your desired institution. If you are their top choice they will, most likely, meet or exceed the salary offer you have already received.

For the vast majority of faculty candidates there will be some, but likely minimal, flexibility in the salary amount that you are offered. Your best approach is to try to gain your own understanding of a reasonable salary offer, either by searching some of the salary surveys that are published, or by talking to your peers who may also be considering a faculty position, or possibly by talking to other faculty either at the institution you are considering or other institutions. Regarding publications, a good place to start is the *Chronicle of Higher Education* and *Academe*, the magazine published by the American Association of University Professors (AAUP), both of which publish annual surveys of academic salaries. Also, you can get more detailed information by questioning other faculty members in the department in which you will work, particularly those in your subject area group, regarding a reasonable salary offer. Keep in mind that at public institutions salary information is considered public information and is available to anyone upon request, so there will likely be little or minimal reluctance to other faculty members sharing their thoughts with you on a reasonable salary offer.

When considering your salary offer, you need to be aware of the time base. The majority of academic institutions will present an offer for a nine-month time period, covering the academic year, which generally starts around the middle of August and extends to the middle of the following May. This leaves the three-month period extending from the middle of May to the middle of August unpaid. In order to provide for income during the summer months, the institution generally pays the nine-month salary over a 12-month period. This may, initially, appear to be a disadvantage of an academic career, but can actually be an advantage since it is possible for you to obtain "summer salary" by a variety of means. This process, in the long run, can actually result in an improved salary.

At some institutions the nine-month period during which salary is paid may be less than nine months, or longer, and the period can extend up to the 12-month calendar year. Although the nine-month academic year period is the most common, the actual time base can vary from institution to institution, and from discipline to discipline. Also, at some institutions it is common practice for the institution to guarantee only a specified percentage of the salary (e.g., 90%) that will be paid by the institution. The faculty member is expected to obtain the remainder of the salary (e.g., 10%) from sponsored research grants. That is, the faculty member is expected to obtain, from external grants, a fraction of their own salary. This situation is very common in the medical field, but is relatively uncommon in the physical sciences, mathematics, and engineering. Nonetheless, it does occur in some institutions and you should address the issue in your discussions with the department head or chair. In some disciplines, such as the medical field, it is not uncommon for the institution to expect the faculty member to raise their entire salary from external research grants.

For faculty members employed under nine-month salary arrangements, the three-month summer period offers opportunities to increase their salary. Research-oriented faculty members can gain salary for this period by working on sponsored research grants. In this case, a portion of their time will be included in the budget in the research proposal that is submitted to the grant-funding agency. This mechanism is an accepted expense by the grant-funding agencies, and faculty routinely will include a portion of their summer salary in the budget of the proposals that they write and submit for funding. Some agencies will permit the entire three-month summer period to be included, while other agencies place a limit on summer salary. The National Science Foundation, for example, permits only one month per year of summer salary to be included on a given grant, with a limit of two summer months of salary for multiple grants sponsored by the NSF, in a given year. This means that if a researcher has three or more NSF grants they are able to charge only a total of two summer months to the grants, no matter the amount of funding for summer support included in the grants.

The US Department of Defense, conversely, places no limit on the number of months of summer salary that may be included in a proposal budget. However, when developing budgets for research proposals, it is important to recognize that most research funding opportunities, no matter the funding agency, will have a maximum dollar amount available for a proposal and this maximum amount is generally indicated in the Call for Proposals or research opportunity announcement. Therefore, funding that is included for items such as summer salary in the budget, will limit the amount of funding available for other items, such as student researcher salary support, travel, materials and supplies, etc. This consideration, in effect, will place a practical limit upon the amount of summer salary support that can be included in a specific proposal. However, it is not uncommon for an active faculty researcher to have successfully obtained multiple research grants in a given year, often from several funding agencies, and that the entire three-month summer salary is paid. In this case, their summer salary can dramatically increase since it will consist of both the summer portion of their 12-month academic salary plus the three months of their salary paid from the research grant funding. You also should be aware that certain academic departments and colleges will limit the time they permit faculty members to devote to sponsored research during the summer months. There are a number of reasons for this, which will be discussed in more detail in Chapter 8, but the policy will obviously limit the amount of summer salary support that universities permit faculty to receive. However, it is fairly universal to permit at least two months of summer time to be devoted to sponsored research activities.

Although the three-month summer period offers a great opportunity to get established in research, the initial one- to three-year period can be frustrating and it may take more time than anticipated to obtain the first grant, particularly from US government research grant funding agencies. Therefore, summer salary support for this initial period of time is critical and should be included in the start-up package. It is common for universities to provide some level of summer salary support to new faculty members so that they can perform initial research and prepare a research proposal for submission to a grant funding agency. Generally, the university will provide a new faculty member with one or two months of summer salary support, over the initial two-year or three-year period, with the exact details to be negotiated. For example, a common start-up package will include a commitment to provide two months of summer support for the first two years of employment. If the new faculty member has a high level of confidence that they will be able to obtain summer salary support from other sources, they may wish to negotiate and trade either one or both of the second year's summer salary support for graduate student researcher salary stipend support, or some other expenditure that will enhance their research progress and development. Salary support for graduate research assistants is very important and these students will be very instrumental in your advancement in research. This issue will be discussed in more detail later in this section.

3.6.2 Travel

Funds for travel expenses may not seem to be a high-priority item to be negotiated in a start-up package, especially for someone just beginning their academic career. However, the situation is actually just the opposite and it's very much worth your time and effort to include travel support in your start-up negotiations, and as early as possible in the process. Travel is a very important, necessary, and fundamental factor in the process of initiating and building an academic research program. In fact, it's extremely difficult to effectively build a research program if one attempts to focus all their time and energy within their home institution, with limited or no plans to travel. In particular, you will need to travel to visit program managers and program directors in funding agencies. You need to build relationships with these people and to establish your credibility with them in order to be successful in getting your research proposals funded. Also, you will need to travel to professional conferences and workshops in order both to present the results of your research and to discuss research trends and developments with your colleagues, and to work on building your network of colleagues and potential collaborators. Travel is, in fact, one of the most important factors and a main building block in your career development.

If you've had experience working in a company or industrial organization, or you've been employed in a government institution, you know travel is at times required, but the travel, housing, and subsistence funds are provided by the home organization. In order to engage in a business-related trip, all that is generally required is approval or authorization from your supervisor to perform the travel. For the most part, you do not need to be concerned with obtaining the travel support funds, and funding amounts and details will be included within the department or organization budget, and managed by your supervisor or department administration. In fact, in most industrial and government organizations the travel details and arrangements will be organized and made for you by the department administration. All you need to do is show up in time to make the trip. However, the situation in a university environment is much different, and the department or college, in most cases, provides limited or no funding for travel on a continuing and routine basis, no matter the purpose, reason, or need for the travel. While university department administrators expect their faculty members to travel to accomplish their goals, participate in their professional activities, and make progress in their careers, they do not, in general, provide the necessary support funds.

Faculty members are expected not only to obtain their own funding for travel expenses, but also in most institutions to make their own travel arrangements. Although there are a variety of sources available to a faculty member for travel funds, the main sources are research grants and discretionary funds under the control of the faculty member or, possibly, the department head or dean of the college. Travel is an expected and permitted expense to include in a research proposal budget. In fact, travel to research program reviews is often a requirement of a US government funding agency, particularly for contracts and grants from mission agencies such as the US Department of Defense, the US Department of Energy, NASA, etc., and they permit travel funds for this purpose to be included in the grant budget. The various divisions in the directorates of the National Science Foundation often require their Principal Investigators to travel for Division program reviews, but they generally provide a grant to a selected university to organize the meeting and provide for travel support, etc. However, this type of arrangement, where travel funds are not included in the grant budget, is essentially limited to the NSF. For industrially funded research, it is often even more important to travel to the supporting organization to give progress reports and to interact with the industrial scientists and engineers. Industrial researchers need to know that their funds are being well spent and that you are producing results that are of interest to them and that you are working on problems that have significance to their projects. The best way to approach this issue is through person-to-person communication and technical discussions, and visits to the organization providing the support funds. Establishing long-time relationships can yield very positive results from a variety of perspectives, but particularly from a continuing funding support basis.

In proposals that you submit to funding agencies you need to include travel expenses in the proposal budget for trips you expect to take. You will need to estimate the number of trips, and the amount of funds for each trip, and also provide a justification for the trip. For this reason, you need to carefully anticipate your travel requirements, as any change to the indicated travel plans will need to be approved, usually by your department administration following the appropriate university policies, and sometimes by the funding agency. Formal changes to grant budgets that require agency approval can sometimes require significant time, on the order of days and up to several weeks, which can create difficulties in travel planning. However, there is minimal restriction on travel, other than the requirement that the travel be in support of and related to the work described in the grant award. Travel funds for visits to the supporting organization and for conference and workshop attendance are common to include in proposal budgets. Travel can be domestic or international, and overseas travel to participate in an international conference, workshop, or meeting is often supported on research grants.

As previously indicated, the NSF periodically sponsors grantee meetings, which their sponsored researchers are expected to attend, and where they present their research results and latest developments to their peers and NSF program directors. These meetings are operated similar to a technical conference, and serve as a forum for performance reviews and the NSF program directors use the researcher presentations as a basis to evaluate the progress being made on the research grant. Often, student researchers under the direction of the NSF supported faculty member are also expected to attend these meetings and to present the results of their research. However, these review meetings are separately funded by the NSF by means of a grant to a hosting university, and participating researchers are not expected to use travel funds from their NSF grants to travel to the review meeting. There is no need for the faculty member to include travel funds to attend an NSF grantees meeting in a research grant proposal submitted to the NSF. The participants will have their grantee meeting travel expenses reimbursed from the holder of the grant to which the NSF has provided the review meeting support funds. Travel funds that are included in an NSF research program grant are generally used for the researcher to attend a professional conference or workshop in their expertise field in order to present the progress and results of their research. These presentations are reported to the NSF and used by the NSF program directors as evidence of progress being achieved in the research program. Travel funds included in an NSF grant are also often used for the researcher to visit the NSF and discuss their research and other subjects with NSF program directors, or to participate in a workshop or meeting held at NSF. Mission agencies such as the US Department of Defense (mainly DARPA) and the US Department of Energy (mainly ARPA-E) also host conference-style research review meetings, which their researchers are expected to attend and present their research results. However, the travel funds for attendance at these meetings are included in the travel budget included in the grant or contract, so before submitting a proposal to one of these agencies, it's best to clarify with the program manager what travel is expected.

Although research grants will provide the majority of travel funds used by a faculty member, other sources of funds are also available. For example, many successful faculty members will manage and spend funds included in discretionary accounts. These accounts are generally not known about, or are mysterious to new faculty members in the first few years of their employment. The discretionary accounts are generally not widely discussed or advertised. However, virtually all universities permit faculty members to establish and manage these discretionary accounts, and use the funds for support of their academic and professional activities. The discretionary accounts are usually established by and operated within the university's foundations organization that operates fund-raising functions for support of the university. The foundation is generally organizationally located outside the university structure, but is operated in close coordination with the university, and for support of the university. The foundations organization will actually own the accounts and the funds contained in them, and will perform all management, accounting, tax reporting, legal, etc., functions. The funds in these accounts consist of "unrestricted gifts" made to the university and the funds originate from the faculty member's fund-raising activities, usually from industrial and business collaborators and supporters. The "unrestricted gifts" statement means that there can be no requirement by the provider of the funds that the funds be used for any specific purpose or project, including any specific purpose identified by the source of the funds. The funds are strictly a "gift." In this regard there are minimal restrictions on what the funds can be used to support. The only real restriction is that the funds be used for university-related business or activities. For this reason, the discretionary account is an extremely valuable asset for research faculty. The funds can be used in support of their other research programs, and for a wide variety of purposes.

Discretionary funds are often used as "gap funding" to keep a research area functioning when funding on one grant terminates and before funding on a follow-up grant begins. The funds may be used for student stipend support, purchase of materials and supplies, purchase of equipment items, payment of page charges for professional publications, support for travel, and any other purpose related to research and professional development activities. These accounts are extremely useful and highly valued. Generally, the funds within these accounts are not subject to "overhead" or "indirect" charges, which means that the funds are very efficiently used for program activities, and a small amount of discretionary funds can have a significant and major impact upon a faculty member's research and professional program development efforts. Certain universities have a policy of "taxing" the discretionary accounts, and a small amount of the discretionary account will be diverted annually for university use. This practice has increased as the discretionary account use has expanded in recent years.

The obvious question is, how does one get a discretionary account established and how are the funds obtained? The funds will usually come from sources outside the university that are willing to provide you with money to support your activities. How do you establish these relationships? Good question! Basically, you need to get someone, usually from an industrial or commercial organization to give you money, and to send funds (a check) to the appropriate university office, and indicate that the funds are "a gift for support of Professor X's (your name) research." The funds will then be delivered to the university's foundation office and used to establish a discretionary account assigned to you for your use.

How do you get an industrial organization to provide the check? The best way is to build professional relationships with industrial scientists and engineers working in research areas similar to your own. If you are making good progress and have novel ideas, the industrial scientists and engineers often will be eager to collaborate with you and your research group, and to hire your students. Sometimes these organizations will support academic research and you will be able to obtain a research grant through the standard research proposal mechanisms. In this case, the normal research grant procedures and policies will apply. However, often the industrial scientists and engineers will want to collaborate with you, or they simply want to support your activities, but they have limited funds in their budgets available for external support. In these instances, they may be willing to provide a small amount of funds to you as a "gift." The funds can be sent to the university and used to establish a discretionary account. In fact, this mechanism is quite common, and a significant number of faculty members have one or more discretionary accounts that they control and manage. Faculty members that have these discretionary accounts find it very difficult to efficiently function without them, as access to the funds dramatically increases their flexibility and ability to respond to events, such as requests to give invited talks, to visit an industrial organization, to attend a professional meeting not related to their research grants, or to visit a US grant funding agency or organization. The discretionary fund accounts are becoming increasingly important to faculty researchers as the research budgets of US government agencies become more limited and more restrictive regarding how the funds can be used.

Why is this issue being discussed under the "travel" subsection as a start-up package item? As a new faculty member, as indicated, you are going to need to travel early in your career, and during your first year. You need to make contact with US government funding agency program managers and program directors. Most likely, you will need to make at least one visit to the National Science Foundation, at a minimum. Also, you should make plans to attend the main professional society conference or workshop that addresses your field of research. In order to make these trips, you need travel support funds, and this needs to be discussed with your department head or chair during the job negotiation process. Travel support funds should be included in the start-up package that you negotiate. A standard offer will generally include funds to support travel for at least two trips, in each of the first two years of employment. If you feel more travel will be required, you can negotiate this with the department head and have the funds included as part of your start-up package. The funds can be provided for the travel support directly, or an alternate method is to request that a discretionary account be established for you. The account does not need to hold a large amount of funds and the exact amount can be negotiated, but an amount in the range of $5000 to $10 000 per year should be sufficient to support your first travel requirements. By establishing the discretionary account, you will be afforded the greatest degree of flexibility. Funds for other items you negotiate in the start-up package could also be provided by means of the discretionary account. Again, your level of flexibility is increased.

3.6.3 Student Tuition and Stipend Support

One of the most important actions you need to address in your first year of academic employment is recruiting graduate students to perform research under your direction. The first students you recruit and employ will be instrumental in building the foundation for your research program, and their efforts will have significant and long-term impacts upon your career. You will soon realize that highly qualified graduate students are in demand and you will be in competition with other faculty members to recruit and start working with graduate students who will assist you in developing your research program. This is, in fact, the most important part of your initial efforts to build your research program. One of the major factors in graduate student recruitment is the ability to offer financial support. In the modern era it is common for new graduate students to be offered some form of financial assistance, generally as a Teaching Assistantship, or a Research Assistantship. Students may also be supported on a Fellowship, which provides funds for their tuition and stipend. Fellowship students are highly recruited and desired by faculty members since the students are essentially self-supporting from the Fellowship funds.

The most common student support mechanism in a university is a Teaching Assistantship (TA), which is offered by the department and makes use of department funds. Students supported on a TA are usually required to perform some duties for the department, such as grading papers in a course, or some other activities in support of the teaching mission. However, a student supported on a TA will be assigned to a faculty advisor, and many faculty members make use of students supported on a TA to perform research under their direction. Also, it's common for TA support to be awarded for partial time, and one-quarter time, one-half time, or full-time financial support may be offered. The partial-time concept permits departments to accept and enroll more students than they could normally support under full-time financial assistance. The TA will also, in general, provide for tuition remission, which means that the student is not required to pay tuition. From the student's perspective the financial support available from a TA is limited, and less desirable than that available from a Research Assistantship (RA), and the RA is more attractive and a superior recruitment entice-ment than the TA. With an RA the student will work directly for a specific professor and the RA support funds will be obtained from research grants that the faculty member has obtained. The faculty member will direct and manage the students that are in his or her program and supported from their research grants. It is also common for a faculty researcher to have a group of students working on research projects that are supported on a combination of RA, TA, and possibly Fellowship funds.

Support for at least one, and possibly two graduate student researchers should be included in the start-up package. The students will work for you as Research Assistants, and the department will provide their stipend support funds. The department should also provide their tuition funds, which may be through remission or by direct payment. The student support should be provided for a two- or three-year period, which will give you the time to get your research program at your new location initiated. As you obtain your initial research grants, student funds may be transitioned to the research grant, or you may be able to use the new funds to recruit and hire additional students. These issues should be discussed with your department head or chair.

3.6.4 Access to Laboratory and Computing Facilities

Depending upon your discipline and research interests and topics, you may require your own laboratory or specialized equipment. If your work is primarily theoretical, and involves simulation or modeling, you may require access to special computing facilities, such as supercomputer time use and support equipment. Desktop and laptop computers and printers for general use are generally readily available in universities and will, most often, be assigned to you for your use. You should clarify the department's policy regarding providing faculty computers, but this will normally not need to be negotiated in the start-up package, unless a computer is not normally provided to new faculty. However, if you require a specialized computer, or access to a supercomputer, you should discuss your needs with the department head or chair, and make sure appropriate funds are included for an initial time period in the start-up package.

If your research is primarily experimental, you will likely require laboratory space and appropriate equipment, as well as materials and supplies. Keep in mind that space is a highly valued, but generally limited, asset on a university campus. It's rare that unoccupied space exists within an academic department, and when vacant space occurs, it's usually quickly reassigned to another faculty member. Therefore, negotiating access to appropriate laboratory space is a high-priority item for your start-up negotiation. The ideal situation is for you to be assigned your own laboratory. However, if no laboratory space for your sole use is available, you may need to share laboratory space with a faculty colleague. In fact, this is quite common and many, probably the majority, of research faculty members work in shared laboratories. This situation is particularly common for research in areas such as microelectronics, nanotechnology, biology, materials science, etc., where large and expensive machines and equipment, and/or cleanrooms are required. Often dedicated laboratories will exist for performing research in these areas and will be managed by professional laboratory managers, technicians, and engineers, with specialized training in the operation and maintenance of the equipment. They will also be trained in safety procedures, and will make sure that all personnel working in the laboratory are properly trained and follow all safety protocols.

Generally, these shared laboratories will require a user fee in order to make use of the facilities. The user fee may be charged on a per machine basis, or it may be a general fee that is charged to work inside the laboratory. The laboratory access and use fees are legitimate expenses to include on research grant proposals, and research grants and contracts provide the majority of the necessary funding. If you, and your students, require access to the equipment within one or more of these laboratories you will need to be able to pay the user fee. For a new faculty member functioning without a research grant, you will need to negotiate access to appropriate laboratories and have the funds included in your start-up package. The negotiation should also include the cost of materials and supplies that you anticipate you will require in your initial efforts to establish your research program. Access and user fees for a two- or three-year period will provide the time necessary to get research results that can be used to write and submit a successful research proposal.

If your research is in a highly specialized area, or the existing laboratories do not have a particular equipment item that you require or would like, you can include the acquisition of the equipment item in your start-up package negotiation. This approach may prove successful. Many universities maintain special accounts to cover the costs for expensive and specialized equipment their faculty members may require. Sometimes state governments will also have funds available for this purpose, which will generally require the preparation and submission of a proposal to compete for the funds. If your university maintains a fund for specialized and expensive equipment, your department head or chair will, most likely, need to negotiate with the dean to obtain the funds, but this is definitely possible. However, if the equipment item is very expensive, the negotiation may prove challenging, and the department head or chair may not be able to respond favorably. In this case you will need to consider other options for gaining access to the necessary equipment. For example, a nearby and collaborating industrial firm may have the equipment item and may permit you access. Also, many national laboratories have specialized equipment items and they will often work collaboratively with university researchers and provide access to their laboratories to both faculty members and their students. You should discuss this issue with your department head or chair since they may have contacts in appropriate organizations that may provide the necessary access. If such an arrangement can be worked out, you may need to have some funds included in your start-up package to travel to the organization to use the equipment. However, if access to a specialized equipment item or instrumentation is not available, you may need to reconsider your research needs. While department heads and chairs make great attempts to satisfy requests, it is not always possible to provide for all requests.

3.6.5 Temporary Reduction in Academic Duties

Initiating and establishing an academic research program is a detailed and time-consuming effort, and an activity to which you will devote many hours of focused attention. An academic career is an occupation consisting of activities that will become your hobby, and you want to be engaged in an activity you truly enjoy. You will find that the time you spend on your professional career extends well beyond the normal nine-to-five day and that you will devote the majority of your waking hours to career-building activities. Your department head or chair very much wants you to succeed as much as you want to succeed. In this regard you have shared goals. However, an academic faculty member is normally engaged in many activities and has many duties that extend well beyond research.

Along with your research activities you will be expected to perform classroom duties and will be assigned courses that you will teach. Teaching duties, in fact, and depending upon your university, may be your primary academic responsibility. Teaching can be a very rewarding experience, and the desire to teach is one of the primary reasons most faculty members embark on an academic career in the first place. However, teaching requires a significant amount of time and effort as you will be required to prepare lecture notes, homework problems, exams, and maybe manage and organize class projects. Grading homework and exams can also require significant time. Often graduate student Teaching Assistants will be assigned to you in order to help in these activities, but you will be required to devote time to managing and directing these student assistants. You will also be required to hold office hours and to work with students having difficulties. These teaching-related duties can, and will, conflict with the time and effort necessary to initiate a research program, particularly during the first few years of your career. For this reason, it is common for academic departments to limit and minimize initial teaching duties for new faculty members attempting to initiate and establish a research program. For example, although the actual teaching duties will vary from institution to institution, many academic departments will assign new faculty members a reduced teaching load, which may consist of one course per year or academic term for the first two or three years. The actual teaching load will vary not only from institution to institution, but also by discipline. Your teaching load should be discussed with your department head or chair during your start-up negotiation, but a reduced teaching load for an initial one- to three-year period of time is a realistic expectation, and your department head or chair is likely to respond favorably to a request for a reduced initial teaching load.

Other factors to minimize the time you will be required to spend on teaching are also possible. For example, if two courses per term are required, you can ask that you be assigned two sections of the same course. This will limit the number of class preparations that will be required, which will help optimize your time use. You may also be able to engage in team-teaching with a colleague, which will permit you to share in the teaching activities with an experienced colleague. This will also help you optimize your time and effort. Another factor to consider is class size and the number of students that enroll in a specific class. Large classes require significantly more time to teach in regard to grading homework and exams, and to directly interact with students. Also, graduate classes generally are attractive for beginning faculty members since the topic material is often similar to their recent interests and students are usually highly motivated to learn the material.

In some cases, a new faculty member will have their teaching assignment waived for the first, and possibly second or third, year. While this will permit the maximum time and effort to be devoted to establishing a research program, in general being completely removed from teaching, even for one or two years, is not necessarily a good idea. Your best opportunities for exposure to students arise from classroom activities, and you need exposure to graduate students in order to recruit them to your research program. The ideal situation is for you to be assigned to teach a course in your research area. This course may be an existing course in the department curriculum, or in many cases, the course may be new and an opportunity for you to develop a course that will expose your students to your research area. This latter approach is very common and used by many successful faculty members as a mechanism both to recruit new students and to begin the training process for students wishing to work in your research area.

Another topic for your start-up negotiation relates to your committee assignments. As previously mentioned, universities are operated with a large number of committees, and you will be assigned to serve on some of them. Committee service is not only necessary for the effective operation of a university, but also provides an excellent venue, particularly for new faculty members, to meet and get to know your department and college colleagues. Committee service is also an excellent method to learn how the department and college function, and to gain an understanding of how decisions are made. This knowledge will make you more effective as your career advances. While you will likely be required to serve on one or more department committees, negotiation of a reduced committee assignment load is a realistic expectation for your first few years. A reduced committee assignment load is a request that is easy for the department head or chair to grant, and in fact, it is very common for new faculty members to be assigned a reduced committee assignment load. Again, it is valuable to have some exposure to committee work since this is a main method to become familiar with the department curriculum and operation. However, by limiting your initial committee assignments to areas that have a direct affect upon your research program development, you can focus your efforts in a valuable and effective manner. For example, you might want to request committee assignments associated with student groups, student recruitment, course and curriculum development, research policy issues, and/or laboratory and computer facilities. This can limit the time you devote to committee work, but provide you exposure to people, issues, and information that will be useful to you in developing your research program.

3.6.6 External Consulting for Pay

Another issue that you may wish to discuss during your start-up negotiation relates to external consulting for pay policies. External consulting for pay is a privilege permitted by virtually all universities, subject to certain policies. You should discuss the appropriate policies with the department head or chair during your negotiation, but keep in mind that the university establishes the consulting policies, and the department head or chair will have little or no ability to change the policy. Therefore, they are unlikely to be able to make any modifications that you may desire. In most instances, faculty members are permitted to spend one day per week, on average, on paid consulting activities. The university policy will generally require that the consulting activity be reported to the university and approved by your supervisor, but there are minimal other restrictions. The university does not define the pay scale, nor are consulting arrangements made through the university. The faculty member consultant is free to charge market rates. The rate of pay may be negotiated between you and the organization seeking your consulting services, or the rate of pay may be fixed by the organization. Often, government agencies and government-affiliated organizations will have defined rates of pay for consulting services that are determined by your credentials, experience, and previous consulting records. These organizations may require you to submit direct evidence of consulting and copies of past invoices or tax records in order to establish your rate of pay. In other cases, particularly for industrial organizations, there is more flexibility regarding payment rates, and in order to determine an appropriate rate it's useful to discuss consulting rates with your department head or chair, or other faculty members who engage in external consulting activities. External consulting for pay is common for university faculty members and can significantly enhance one's income.

3.6.7 Professional Development

Professional development opportunities should also be discussed during your start-up negotiation. For most new faculty members, professional development will include your participation in and attendance at professional conferences and meetings. Professional societies often sponsor seminars and meetings that present issues associated with research developments and trends, as well as annual discussions of US government granting agency budgets, trends, and policies. Professional development may also include visits to research funding agencies, industrial organizations, and visits to other universities to discuss research topics and potential collaboration with your colleagues. Many of these activities were discussed earlier in this chapter under "travel," and adequate support funds should be negotiated and provided in the start-up package. Professional development may also include additional educational activities, such as tuition for courses you may feel appropriate and necessary for your career development. Again, unless the professional development activity is offered to all faculty members on a routine and regular basis, any item that requires funding needs to be negotiated and defined in the start-up package.

3.6.8 Dual Career Opportunities

It is becoming increasingly common for a married faculty member being recruited to have a spouse who is also in the job market. The spouse may also be seeking an academic faculty or staff position, or they may be seeking a suitable position in a commercial or industrial organization. If the university to which you are being recruited is located within or near a major urban area, there may be good spousal employment opportunities. However, many universities and colleges are often located in rural areas and the university or college may, in fact, be the major source of employment in the area. Many universities today recognize that, in order to successfully compete for highly qualified faculty members, they can gain a significant recruitment advantage by offering a spousal employment assist program. This can range from simple assistance in identifying a suitable employment opportunity, either within the university or at an outside firm located in the nearby area, to a formal employment program. Often the hiring university will have staff members who are trained to assist the spouse in finding suitable employment. The university may also be prepared to offer the spouse a job, commensurate with their skills, training, and experience. For example, if the spouse is also looking for a faculty position the department head or dean may be willing to approach another department with a faculty opening and help negotiate an employment offer for the spouse. This situation occurs most often when the university is searching for an upper administrator and, in order to hire their top choice, they need to also find suitable employment for the spouse. However, there are also numerous instances where a university is willing to hire a new and young faculty member, as well as their spouse. Most often, when this occurs, the dean of the college will be involved, and the dean may need to move a new faculty position into the department that will hire the spouse. If spousal employment is a major concern and issue in your recruitment, you should discuss the situation with the department head or chair, and question them regarding spousal employment assistance programs.

3.7 What We've Learned

Your academic career begins with the recruitment process. It's extremely important to understand this process and to negotiate the resources that will permit you to be successful. Once you successfully seek and accept an academic faculty position, the tenure-track clock starts, and you embark upon a professional career with well-defined timelines. It's critical to get a good start, particularly in establishing a research program, but also a good start on the other duties of a faculty member. Faculty members are evaluated based upon their performance in three areas: teaching, research, and service. Of these, becoming established in research is the most challenging, and requires the most time. However, this is only one of the three areas in which performance will be evaluated, and the other two areas must not be neglected.

The most challenging aspect of establishing a productive research program is obtaining the necessary financial resources to obtain and support the research. This includes recruiting and providing support to students, post-doctoral assistants, and other necessary support staff. You must also acquire the resources to secure necessary equipment and instrumentation, access to laboratory facilities, materials, and any other items necessary to perform the research. Negotiating an appropriate and adequate start-up package that will provide funding for items and activities necessary to initiate an externally funded research program facilitates this process. You will need to be able to focus a very significant amount of your time on research, but without neglecting your duties in the teaching and service areas. Therefore, the items to include and request in your start-up package should be carefully considered. These issues should be carefully discussed with the department head or chair, and before your final acceptance of a faculty member position. It's most important to get a quick and positive start to your new career.

4   Getting Started and Marketing Your Research

Now that you have gone through the recruitment process and have successfully found your dream job as a university professor, you're ready to begin your career. You'll soon discover that you're about to make a major transition in the way you experience the employer–employee relationship. You most likely have been in the mode of operation where you have had a boss or manager to whom you have reported, and this person had major responsibility for defining and determining the tasks and projects upon which you devoted your time and energy. No matter if you have only recently graduated with your PhD, or if you have worked in industry or government, you have been managed by someone else who has provided you with guidance and assigned the duties you were expected to perform. In many respects, this is a very comfortable situation in that your time and energy have been directed towards applying your skills and knowledge to tasks and projects essentially defined by others, and your main effort has been directed towards completing these tasks and projects in a timely and effective manner. You have likely had numerous routine meetings with your manager or thesis advisor, and may have interacted with him or her on a regular basis, which could range from multiple daily meetings to weekly or monthly communications and meetings. You most likely have also attended regularly scheduled staff meetings, which probably occurred on a weekly basis, or possibly more often. Your performance evaluations have been determined by how effectively you have completed your assignments and, of course, on the quality of the work you performed. You have likely put much of your own ideas and initiative to work in performing your tasks, and you may have had, in fact, much input into the definition of your projects. However, the final definition of your tasks and projects has been the responsibility of your manager. No matter the process by which your tasks and projects have been determined and defined, you have been successful, as evidenced by your recruitment and hiring into your present academic faculty position.

As you transition into your new occupation you will soon learn that these very clear and well-defined lines that exist in the normal employer–employee relationship are blurred and not at all clear for faculty members. For example, your manager in an academic department is your department head or department chair, and the person who occupies this position will be responsible for assigning you your academic duties and conducting your annual performance evaluation and review. However, you'll soon discover that your academic peers and colleagues also have significant input into your performance evaluations and are important voices in your potential promotion and advancement in your career. You want to have friendly and collaborative relationships with them. However, your actual day-to-day duties are only generally defined, and other than meeting and teaching your assigned classes and holding office hours, serving on any committees to which you are assigned, managing your research activities and students, etc., you'll discover that you will, most likely, have significant freedom and time to devote to other activities. You will likely experience limited interaction with the department management. For example, you may go weeks without actually personally meeting or even talking to your department head or department chair. Most academic departments schedule regular faculty and staff meetings, but these generally occur only once or twice a month, and the interactions between the faculty and staff members and the department leadership is in a group environment not appropriate for discussions regarding personal matters. Of course, you may request and schedule a meeting with your department head or department chair at any mutually convenient time. However, most of these meetings will occur at your request, unless there is a particular reason your department head wishes to meet with you individually.

You'll discover that you have independence and freedom and that you personally will determine how much time and effort you devote to any particular activity. You will also discover that there will be many requests for your time and involvement in projects and events that will distract you from your main goals. Time management will become a major factor in your daily routine. This can, however, become a problem if not well controlled, and effective time management is probably one of the most difficult challenges for new and young faculty members to master. With no one actually directing you on a daily basis and with the many and diverse distractions that occur on a regular and continuing basis in an academic career, keeping your efforts focused on priority projects and goals can be a difficult challenge. You will discover you need to carefully evaluate and consider every request for your valuable time. Unfortunately, you will not be able to respond positively to every request. Everyone, and particularly new faculty members, face this challenge and must learn to effectively manage their time in order to be successful. Often, it becomes necessary to decline to participate in certain projects or functions if your participation will require significant time that would divert you from your main goals.

4.1 The Professor as an Entrepreneur

As indicated throughout this book you need to immediately focus your efforts upon initiating and building your research program. This effort needs to become a first priority. In order to effectively build an academic research program, you need to realize that you are beginning a career where you will essentially function as an entrepreneur. This is one of the major challenges faced by all new and young faculty, particularly those just graduating with their PhDs, but also those transitioning from a career in industry, the private sector, or government employment. There is good news and bad news regarding this transition. The good news is that you will have extensive freedom and independence in pursuing your interests and building your career. You will have the freedom to pursue essentially any research area in which you have interests and feel you can make significant contributions. This freedom and independence is both exhilarating and intoxicating. It's probably one of the most attractive characteristics of an academic career. The bad news is that you will need to obtain, on your own initiative, the resources required to establish your program. The primary requirement is, of course, you will need to obtain from external sources the funds required to build, operate, and maintain a research program. If you are joining an established group with existing laboratory facilities, your transition will be facilitated and made much easier. However, many new faculty members will not have this advantage. Your university, of course, wants you to succeed and will actively work to provide mentoring assistance and help you in getting established. They will provide you with office and laboratory space and other support services, but these support efforts are always limited. The main effort to establish your research program falls squarely upon your shoulders, and you will need considerable focus and determination, as well as creativity, to successfully achieve your goals.

You will quickly find that your efforts to build a funded research program are much like the efforts and skills required in starting a new small business. You will essentially operate as an independent entrepreneur within the university and you will be establishing a business that will function in essentially the same manner as any conventional business. That is, you'll need to identify your products (i.e., your research and students... and yes, in this context students are like products in that they will be recruited for employment by industrial organizations, government agencies, academic institutions, and other organizations). In fact, the students you mentor, train, and graduate will be a very important factor in establishing your professional recognition. You will also need to determine a market for your products; a market that consists of the funding agencies you approach for support, the industrial and business firms with which you will collaborate, and the professional community in which you will contribute your expertise and research results and provide your volunteer services. You'll need to recruit and hire students that you will mentor and advise, and direct in performing the research you pursue, and you'll also need to publicize your research, in the form of professional publications and presentations at technical conferences and workshops, presentations to prospective funding agencies, invited presentations at peer institutions, and other professional venues.

By the way, both your research results and the students you graduate represent powerful advertising for your research "business." This serves as the "marketing" dimension to your business. As your research successfully progresses and you publish your results, you'll find that your students may become highly recruited, particularly if your research is in an area of high current interest, and this will help you to build relationships with industrial, business, or government organizations and funding agencies that are interested in your research. As your students earn their degrees and graduate and transition to their own careers, their success will directly affect your reputation. You'll find that you start receiving telephone calls and other communications from personnel at industrial organizations, other universities, and government institutions who are in the process of recruiting new employees and consider your students as potential employees. They will question you regarding the performance of students nearing their graduation date and request recommendations for potential students that are seeking employment, or may be a good match to their employment needs. They will often also request the names of students working in your research group that may be suitable for future employment, although the students' graduation may be two or three years in the future. They will follow the progress of these students as potential future hires. Occasionally, as you become established, some of these recruiters may have previously been your own students. This is, of course, a strong indication of the success of your program, and a source of pride for you and your research program. Recruiting high-quality graduate students is probably your most important and immediate priority, and a priority that will exist throughout your career. You'll find you need to invest significant time and effort on this activity. However, you'll also learn that your investment of time, energy, and effort on graduate student recruitment activities is well worth your attention and, in fact, is a very enjoyable and enlightening experience. Working with graduate students is one of the high rewards of an academic career. The payoff in time and effort, in the form of successful recruitment of quality graduate students, is immense and difficult to understate.

4.2 Getting Started

So, how do you actually get started? You need to initiate your research activities as soon as possible, and this needs to be your initial focus and primary goal. There are several ways to do this. First, if you are already performing research, you should continue your current activities. This may involve interacting and continuing to collaborate with anyone you have been working with at your previous institution or organization as this can serve as a good springboard to your new career. Second, your start-up package negotiated during your recruitment can serve to provide an initial source of funds to recruit one or two students to work with you. Your focus here should be to do sufficient research to obtain initial results that you can use to prepare and submit a proposal to a funding agency for more substantial financial research support. Third, your university may offer some sort of "seed" funding opportunities, generally on an annual basis, that are intended to assist new faculty members in getting started in research.

These seed funding opportunities are in addition to, and not the same as the funds provided as part of your start-up package in the faculty member offer discussed in Chapter 3. Often, universities will allocate a fraction of the overhead or indirect costs collected from funded grants and direct it for the purpose of initiating new research, primarily for new and young faculty, but also for existing faculty members wishing to change their research areas, and also for new programs that the university wishes to initiate. Other sources of seed funding derive from money raised by foundations and other sources associated with, but external to the university, and obtained through the efforts of university fund raising activities to build the college or university endowment. This effort has always been a significant activity for private academic institutions, but has also become increasingly important for state-supported institutions in recent times, particularly as operating budgets have experienced limitations and, in some cases, reductions. Focused and directed effort to build endowments has become a major component of university financial planning, and these activities involve significant energy direc-ted towards establishing the financial resources required to build a robust and stable academic and research environment. Virtually all colleges and universities have foundations that solicit financial donations on a routine basis from sources external to the university, such as alumni, industrial organizations, community associations, and other supporters.

Opportunities for seed grant initiation funding may be offered on the department, college, or university level, and are generally advertised by the college and/or university research offices. The total funding will generally be very limited in amount, and the level of funding available for a specific proposal will be modest and made available for a limited period of performance. The main purpose of these funds is to provide the proposer with some resources to obtain background or initial results that can be helpful in supporting and developing ideas for proposals to be submitted to funding agencies for more substantial support. However, you'll definitely want to take advantage of these opportunities since the support, even if limited, is often sufficient to gain some initial results. University leadership recognizes that providing a limited amount of seed funding can be an important and significant factor in helping new and young faculty, or more advanced faculty in the process of changing their research direction, to formulate and initiate their new research. Initial and preliminary research results are very important in preparing initial grant proposals for submission to funding agencies.

No matter what mechanism you use to begin your research activities, your focus needs to be centered upon obtaining initial research results that can be used to support requests for further research support. You need to be able to make as strong an argument as possible that your research is addressing an important problem, and that your research and your research approach offer a possible solution, and one that will advance the science or technology. Funding agencies, in general, are reluctant to fund research proposals that do not include initial results which provide an indication that the research will prove successful, and funding agency program managers do not, in general, like to fund research that is highly speculative or not supported with adequate scientific or engineering support data. This is becoming more important, particularly in recent years, as funding agency research budgets stagnate or are reduced. For scientific and engineering research, in particular, program managers like to see preliminary experimental data and measurements, or initial theoretical model predictions and simulations that support the main objectives of the research being proposed. They want to see some sort of evidence that their limited research grant funds will be well spent on a new or promising approach to a problem with which they are well acquainted.

Inclusion of preliminary measurement or simulation data is very important and can, in many cases, make the difference between the program manager accepting or declining a proposal. Preliminary data can also be obtained from literature reviews, or obtained from collaborators in industry or elsewhere. Of course, all data and information from external sources needs to be properly credited and referenced, but such information can also be used to build the case for the research that is being proposed. Previous research that has been performed, research conducted with start-up package support, and results obtained from the seed funding opportunities offered by your college or university provide the preliminary results and data that can be used in a more detailed proposal to be submitted to an external funding agency. The start-up package and seed funds are generally used to support graduate students to perform initial research, help collect data, perform experiments, travel to visit a funding agency, and other expenses related to proposal development activities. The permitted expenses are defined in your start-up package details, and the seed funding announcement will generally describe the allowed expenses permitted for those opportunities.

4.3 Consulting and the Relationship to Research Program Development

Obtaining initial research results and background information for inclusion in a new proposal can be a daunting challenge for new researchers, particularly if the proposal is directed towards a new area or subject for the researcher. The start-up package and seed funding opportunities offered by the university help in this regard but, as mentioned, provide limited amounts of financial support. Other approaches are also available. Working and collaborating with industrial and governmental organizations by means of consulting agreements offers one attractive approach, and consulting arrangements can be an important and positive factor in building research support mechanisms. Such consulting arrangements generally expose you to "real world" problems that can be the source of data or results that can be important in writing future proposals. Often, this type of information can help make the difference between success and non-success in obtaining grant funding, particularly if the information results in a new and novel approach to an existing technical problem. Government program managers always welcome novel approaches to existing problems and they are generally well aware of problems that are limiting or degrading a particular technology in current use. Consulting also helps you build relationships with industrial scientists and engineers, and these relationships can endure over significant lengths of time, sometimes extending over decades. As a personal note, I have worked with certain industrial scientists and engineers over decades, where they and I have both changed the institution where we were employed several times. The names and faces remain the same, but the employment location changes. These long-standing relationships result from collaboration that develops into friendships on both the professional as well as personal level.

These consulting relationships also can prove important for certain grant funding opportunities where the funding agency requires industrial interest in the research you propose. This is becoming increasingly important for certain grant solicitations, particularly those offered by the NSF, DOD, DOE, and NASA, and your relationships with industrial scientists and engineers can facilitate your ability to respond favorably to the funding opportunity. Generally, these research proposal solicitations require a support letter from the industrial organization, indicating their interest in your work and their desire to support your research with some level of financial or other resources. The support often consists primarily of in-kind resources where you are granted some level of collaborative relationship, but also may consist of some level of financial support. The in-kind support can prove valuable and may offer you, and your students, access to industrial laboratory facilities that you don't have in your home institution, or it may offer you access to experimental devices or products that are in development, or development data that assist you in your research activities. This access and data can be very important in helping you become aware of significant problems that limit current technology, and this information can significantly affect your research directions.

Research directed towards developing solutions to real world problems experienced by industrial developers is often highly valued both by the industrial organizations and also by government funding agencies and the program managers who are interested in the technology under development. The consulting collaboration can also provide the initial data, or experimental validation and verification of ideas that provide support for concepts that you plan to include in formal proposals to be submitted for grant funding. The consulting collaboration will generally significantly enhance your ability to perform the research you propose, and may expand and enhance the scope of the research you are able to perform. And last, but certainly not least, consulting will provide you with additional monetary income that will augment your university salary. You should make an attempt to participate in consulting activities as an important dimension to your research program development, particularly if your research is in engineering or science fields where there are industrial or government organizations interested in your research. Both the information you learn, as well as the personal contacts and relationships you develop, will prove useful in your research program development activities. Your academic institution will have policies that regulate consulting activities but, in general, all universities permit academic faculty consulting privileges that typically amount to one day a week, on average. Your consulting activities will be subject to approval by your department head or department chair, and will be reported in your annual activity report that is required of faculty. You will, in general, most likely find your department head supportive of your consulting activities.

4.4 University Mentoring

As previously indicated, your efforts to build a research program within your university are similar to the activities required to establish a small business. However, your entrepreneurial activities do differ in an important respect from a commercial company in that you work for a university, which provides you with a base for your "operations" that consists of office and laboratory space, as well as some laboratory equipment and supplies, but the university itself is only marginally involved in your efforts in that there is little guidance or assistance in determining research directions or goals. University administration, leadership, and faculty want you to be successful, but they have limited effect upon your research interests or research projects. In essence, they really have little interest in determining what you actually direct your research activities towards, and they primarily want you to become established and to become successful. The university faculty and management, of course, have invested significant time and resources in your recruitment and the support and start-up packages that you have been provided. They clearly expect you to succeed. They also offer morale and logistics support to help in your professional initiation, primarily in your research program activities, but also in helping you become an effective teacher.

Many universities now offer fairly extensive mentoring and training programs concerning research program development, identification of available funding resources, how the university internal grant processing and monitoring system function, the appropriate state and national government regulations that govern your research and professional activities, as well as how to manage your interactions with your students and colleagues. Many universities also offer internal mentoring programs where a new faculty member is "teamed" with a more experienced colleague to help them initiate their research program and to provide advice and guidance regarding teaching methods, student management, etc. Your department head or department chair and faculty colleagues, as well as the college and university administration and leadership, are all invested in your success. You can be assured that everyone wants you to succeed! You should be prepared to take advantage of these good offices. Mentoring options should be discussed in detail with your department head or department chair and your faculty mentor at the earliest possible opportunity. They can provide you with good advice, particularly regarding university policies relative to sponsored research.

4.5 Know Yourself

As a new faculty member you, most likely, have a limited view of yourself from a professional perspective. You may think you have a good understanding of your capabilities and what you want to do, but likely have limited knowledge of how to become established. You need to address the question: How well do you actually know yourself, particularly from a research and academic professional perspective? You should evaluate yourself from the point-of-view of questions such as: Are you primarily theoretically oriented, are you an experimentalist, or do you enjoy both theory and experimental work? Are you someone who enjoys games and puzzles? Are you a person who wants to know how things function? Are you interested in the world around you? Are you a naturally shy person, or outgoing? Do you function effectively in social situations? How well do you communicate, both in person and in writing? Do you like to join volunteer organizations and do you willingly share your talents and time with others? Most importantly, are you an organized person who is able to focus your energy upon a desired task? These are only a few of the questions that can be asked, but indicate the sort of thinking that will define your professional character and your professional career. You'll learn that you need to function effectively on many levels, some scientific, some professional, and some social. The questions can be a bit tricky to answer, but the honest answers will determine how you approach research, and how you attempt to go about building a research program. Complete answers require a thorough self-evaluation not only of what you have accomplished in your research career up to the present time, but also how you approach research and how you address and solve problems that you encounter. Most importantly, you need to determine and define what research problems and directions you want to pursue in the future, and complete answers to the questions will be determined from your past and your experiences up to the present time.

The first step in the process requires that you establish a clear picture and definition of the research you wish to pursue. This step may seem obvious and simple, but is actually much more difficult than it initially appears, for it requires that you not only consider the research that you've been pursuing and with which you are most familiar, but that you also consider the research topics that you wish to pursue on a long-term basis. These research interests are not necessarily the same and, in fact, sometimes differ in a significant manner and may not even be in the same area. You need to have a very good idea of the state-of-the-art in your preferred field of research. Keep in mind that, in order to be successful, you are going to need to spend much time and effort pursuing your career objectives. In essence, your work is going to need to become your hobby. You'll find that you'll be spending the majority of your awake time on your research and academic activities, and you'll soon learn, if you haven't already, that an academic career is most definitely not a typical nine-to-five occupation! It's best if you really enjoy what you're doing and that you look forward to going to work. As an example, I recall when I started my academic career, I would go to my university office to get some work done on a holiday, and have difficulty in finding a parking space in the parking lot. The laboratories and offices were full of graduate students and faculty researchers! Holidays, weekends, etc., were ideal times to get work done since other distractions were at a minimum, and significant attention could be directed towards performing research, generally with good progress achieved. If you don't enjoy what you are doing, the process quickly becomes tedious, and the prospects for success will diminish.

One of the advantages of an academic career is you have a large degree of freedom and latitude in your time and activities, but the downside is that it's easy to get distracted, and for time to pass with little being accomplished. You need to become an excellent time manager and to focus your effort on productive activities. In the current age the need to actually be in your office or laboratory in the university has changed somewhat, particularly for research that can be performed on a computer, such as theoretical modeling and simulation research, or library literature research, since most faculty members now have home offices equipped with computers, printers, etc., connected to their university networks and the internet. Much focused research can now be effectively performed in the home office environment. However, the dedication to the research and the long hours devoted to making progress remain, and only the location where the research is performed varies. Career advancement and success will require that you become fully immersed in the process. You'll be thrilled when you receive your first research grant and will want to spend as much time as possible performing the research. As the research progresses, and you start obtaining good results you'll be even more thrilled when you publish your first paper based upon that research. When this happens, you're on your way to a successful career.

4.6 Your Research Area

In order to get a good start, definition of your research topic needs to be immediately addressed since this is probably the most important issue that needs to be settled. Extending the entrepreneurial activity and start-up business analogy, keep in mind that the vast majority of new start-up businesses fail, and for the same reason: lack of focus! That is, the people starting the business get distracted from their main goals and lose sight of what it is they're trying to accomplish. This is a particular problem in attempting to establish an academic research program since numerous temptations and distractions routinely arise, and these represent forces directed towards pulling one from the main objectives that are being pursued. For a start-up business, the distractions may be financial, and the need to do tasks that are a distraction from the main goal in order to gain some revenue and support funds. While obtaining revenue is obviously very important for a new business, the distraction from the main goal may be counterproductive for the new business and can lead to business failure. This type of distraction is generally not a major problem for an academic researcher since the faculty member's salary does not depend upon income from their research activities, with the exception of salary augmentation from research performed in the summer months and paid for from research grants. However, time and technical distractions can also be a problem, and these can and normally will affect an academic researcher. There are only so many hours in a day, and effort spent on pursuing a distraction reduces the time available to direct towards productive research. There is always a tendency to get pulled from the main goal by side projects that you feel obligated to perform for various reasons, or that may appear initially interesting and in which you'd like to become involved, but require time and effort that should be more appropriately and profitably applied towards efforts that advance the main goal. Since resources and time are always limited, there is a clear need to focus and limit involvement to activities that are in support of your main goals and objectives.

For a new business to be successful the managers and directors need to focus all resources upon making progress toward the main goals and objectives of the business and to direct and limit employee time and effort to projects and tasks that support these activities. In this regard, there is a strong parallel between initiating an academic research career and establishing a new business. Both require identification and establishment of a set of well-defined objectives and goals, and establishing a strong focus upon activities in support of accomplishing the overall goals. The distraction factor can be especially strong in an academic environment since you also have teaching, committee work, and other academic duties associated with your job that will require your attention. Research is actually only one fraction of your duties and you'll need to learn to budget your time to make sure you have adequate time to direct and focus upon your research activities.

Distractions can arise from a variety of sources. For example, you may find yourself contacted by faculty colleagues from both your home department and from other departments, with research interests similar to your own, and who would like to collaborate with you. Keep in mind that you have been recruited through an intense and competitive process, and that your research discipline and the results that you have achieved and reported have played an instrumental role in your selection as the successful candidate. It is only reasonable that other faculty members working in your research area would see you as a potential collaborator. In fact, you may have already discussed potential joint activities with certain faculty members during your recruitment process. Some of these collaboration possibilities may be attractive opportunities and may prove extremely productive. You may find that you're able to leverage your work against that of a more established colleague to your mutual benefit, and this approach may significantly enhance and accelerate your progress towards your career objectives. Under these conditions you should most definitely pursue the collaboration since it's a potentially positive factor towards establishing yourself professionally. However, you may also receive overtures for partnerships from other faculty colleagues working in areas that are either not directly in your research field, or who are pursuing an objective or technique that is not consistent with your research. In these cases, collaboration with these faculty members may well prove to be a distraction from your main objectives, and ultimately prove disruptive to your academic career progress. You should carefully evaluate these opportunities, but if the research collaboration is not consistent with your goals and objectives, you should decline collaboration in these cases. Of course, you may find it very difficult to decline the collaboration, and you'll need to be very collegial and political in your response to the faculty member. You don't want to start off your academic career by alienating a colleague, particularly one who has reached out to you and approached you in a spirit of collaboration.

Also, there are other situations where declination of a joint activity opportunity may prove to be quite complicated. In particular, it may prove uncomfortable to decline to work with someone if the interested faculty member is very well known and professionally established. You may see the person as someone who can introduce you to funding agency program managers, well-known researchers at other universities, industrial scientists, etc. Also, if the person interested in collaboration is a tenured senior faculty member you recognize that they'll have a voice in your own tenure and promotion evaluation. In fact, they may be contacting you with the specific intention of providing you some assistance to help you get started. In this instance you may want to accept their offer to work together, but you'll still want to work out some arrangement where your efforts do not distract you far from your own research goals and objectives. Honest and frank discussions with all faculty members and research scientists who approach you for possible collaboration should be conducted. You should be prepared to openly and honestly discuss your career objectives and where you plan to focus your research efforts. You'll most likely find them to be very open and understanding, and they most likely contacted you in the spirit of helping you to get started. In many cases, if your research objectives are not consistent with those of the faculty member interested in collaboration, they will direct you to more appropriate faculty members with research goals consistent with your own. They may even introduce you to their colleagues who are more appropriate for you, and closer to your area of interest. They may also be able to introduce you to appropriate research funding program managers. Established research faculty members are generally familiar with program managers in various technical fields within funding agencies, and they can likely identify for you the program manager who funds research in your area.

For all the reasons indicated above, it is fundamentally important for you to identify the main goals and objectives of your research and, more generally, of your overall professional and academic career. Establishing an overall career plan is important, and in fact, is required for certain program opportunities. The NSF CAREER program (described in Chapter 7), for example, requires the preparation and presentation of a career plan that addresses not only research goals, but educational and outreach activities as well. Focus is clearly a major issue and is very important in establishing a research career. There will be many opportunities to become involved in activities that may initially appear interesting, particularly if you don't have significant external support at the time, but if these activities are not directed towards topics that are consistent with your career objectives and goals, becoming involved in these activities may actually do more harm than good for your overall advancement. Learning to say "no" at the appropriate time is a very important, but difficult lesson to learn.

4.7 Your Research Area within Your Department

Let's assume that you have tentatively addressed your research topic, and have decided to continue working in the area in which you performed the research that resulted in the award of your PhD. After all, this is the research area and topic with which you are most familiar and upon which you have already spent significant time, generally on the order of three to five years. You know the area and topic very well, and you are, in fact, a world leader on the subject. Also, you have likely already published on the topic in scientific and professional journals and, perhaps, given presentations at professional meetings or conferences. In fact, the majority of new and recently graduated faculty members initially plan to continue their research in the area in which they performed their thesis research. In the recruitment process your research was carefully considered and evaluated, certainly with regard to its quality, but also with consideration to the topic of the research and how the research "fits" into the overall spectrum of research conducted within the department in which you work.

New faculty expertise recruitment decisions are made by the department chair or department head, or hiring committee, generally in consultation with and advice of the department faculty. New faculty positions are generally directed towards specific expertise areas within a department, and defined to address either a void in the spectrum of disciplines within the department, or to augment existing topic areas, or to address new and emerging research areas. Hiring decisions are either based upon these criteria, or simply directed towards recruiting the most qualified new faculty member available regardless of research area. No matter the intent of the hiring process, new faculty are expected to participate in department area "groups" defined by technical discipline. The faculty in most academic departments will self-assemble with their colleagues into technical and scientific groups that pursue similar academic and research interests. They organize and coordinate both their teaching and research efforts. Each area generally has a list of both core and associated courses that provide the necessary education and knowledge for students wishing to pursue their education in the area. You will work in this environment, and for this reason, your research topic will be focused upon a topic that generally "fits" into the technical area in which you were hired, even if your technical subject is only loosely related to the area group in which you are identified. Your teaching activities will generally be associated with courses that support the technical area group, although you may also be assigned courses that are introductory and cover basic subjects required of all students.

Although you will be assigned to a specific technical area group there is a lot of freedom to define and determine exactly what specific research topic you choose to pursue. From a department perspective there is considerable interest and emphasis upon your success, and all that really matters is that you are successful in research program development. What technical area you decide to pursue is actually a secondary consideration from the department perspective. If you decide to pursue a research area that differs from the area for which you earned your PhD, there will be no issue with your decision. In fact, you may find yourself collaborating with faculty in other areas within the department, or within other departments within the university. There is generally no limitation on this, and in fact, such cross-disciplinary research collaborations are encouraged both by the university and also by funding agencies.

4.8 When Should I Change My Research Area?

So why would you want to change your research area from that for which you earned your PhD, and which provided the background and expertise that served as the basis for your recruitment? Why should you not continue to work on this topic? The answer may be, of course, that you, indeed, should continue to work in the same area and on the same topic. This is particularly so if you work in an area that is of significant current scientific or engineering interest and still has a lot of unanswered questions that require further research. In this case, you're well positioned to continue to perform research in this area since you already have a good record of success. However, your research area may be rapidly advancing and maturing, and there may be a large body of researchers already working in the field. If you aren't well established in the field on your own reputation, you'll likely find it difficult to acquire funding. The situation can become complicated for PhD graduates who worked in groups under the direction of well-known and established professors. On the positive side, working with a well-known researcher can open many doors that can present excellent opportunities for you to meet other researchers, funding agents, industrial scientists, etc. These opportunities can serve as useful avenues for you to advance in your career. However, if the research area is scientifically mature, attempting to continue research in the same area or on the same research topic will place you in competition for research funding with more established researchers that may include your own thesis advisor. This can place you at a disadvantage as you attempt to obtain research grant funding, particularly if you haven't had the opportunity, under your own name, to publish papers or present research results at professional conferences and meetings. This issue can be particularly discouraging in a time when the availability of research funding is limited or declining.

When funding agencies experience stagnant or declining budgets, they have a tendency to direct their funds to established researchers in whom they have confidence of good progress being achieved. Of course, many funding agencies have set-aside funds that are available for new and recently graduated researchers. You will want to compete for these funds, but the amount of funding is generally limited and competition for these funds can be intense due to a large number of applicants. Also, if you are working in a technical area where commercialization of products is a goal, and in fact, commercialization is occurring, you may wish to consider your options. When industrial engineers and scientists are working on research similar to your own interests, there are positive and negative issues to consider. First, industrial interest in the research may provide options for collaboration, particularly if your research offers a novel approach. You may find that collaboration with the industrial engineers and scientists provides you with exposure to real-world problems and potentially experimental facilities that you do not have at your home institution. In this case, the research area and topic is probably worth investing your efforts. You are likely to have the opportunity to produce results that are immediately publishable, which will enhance your professional visibility. However, if the research topic is rapidly maturing there may be limited opportunities to perform the basic research that most research funding organizations wish to support. Most funding agency program managers are constantly looking for novel approaches to problems from a basic research perspective. Advancing an established or mature topic is suitable for industrial interest where new devices, components, or systems are in the process of being commercialized, but continued research on such a topic will have a declining interest for sponsored academic research. As these topics advance, further research generally requires higher levels of research and development funding, and this is difficult to sustain in an academic environment.

If these conditions describe your situation you may want to consider an area or research topic where you can provide a new and novel approach or interpretation in a new or emerging research area of interest. Of course, you want to base your research in a topic in which you have knowledge and confidence. However, extending your research to new and emerging areas will position you for the future. This approach is particularly advantageous if the new area is an extension of your current research. For example, my primary research area was in semiconductor electronic devices for microwave and millimeter-wave applications. I worked both in experimental characterization and circuit development, and in modeling and simulation. My original research was focused upon Si devices for microwave and millimeter-wave applications, and the primary research device was the IMPATT diode (an electronic device that is not widely used today). However, new research opportunities developed as the range of device types expanded, including both diode-based devices as well as a variety of transistor structures, and as the semiconductor material from which the device was fabricated transitioned from Si to GaAs, and then InP, and then to a range of compound semiconductor heterostructures such as AlGaAs/GaAs, GaInAs/InP, and then to the wide-bandgap semiconductors including semiconducting diamond, SiC, GaN, and then heterostructures of AlGaN/GaN and others. Each semiconductor material system had advantages and disadvantages for certain device types, and the worldwide effort to develop the most advanced and optimum devices based upon varying material system has been ongoing for well over the past five or six decades. There was a large increase in semiconductor materials-oriented programs as the semiconductor material growth technology evolved and advanced. At the same time the RF electronic devices evolved from the early three-terminal bipolar transistors, to higher performance two-terminal devices such as the IMPATT and Gunn-type devices, and then back to advanced, high-performance three-terminal devices such as scaled bipolar transistors and heterojunction bipolar transistors (HBTs), to a variety of field-effect transistors (MESFETs, HEMTs, and HFETs), and then to high-performance integrated circuits. Devices for low-frequency, high-power applications, and devices for optical applications such as LEDs and solid-state lasers, also were developed. The point is that it was (and remains) a very dynamic field, and it was a relatively simple task to extend research into the new and emerging semiconductor material systems and address new electronic and optical device structures. Understanding of the basic material characteristics and device physics principles were the same, and only required consideration of issues and physical effects that specifically pertained to the new materials systems and specific devices and their dc and RF operating conditions, as well as their circuit environment parameters.

I was able to extend my research and maintain my research career well beyond the research I performed for my PhD thesis. While extending research to include these additional effects and considerations was not necessarily simple or easy, the research was of interest from both a theoretical and practical perspective, and adequate research funding opportunities were available. Researchers in this field, of which I was only one of many, had a very fertile range of topics to pursue in our research and a large number of researchers in the field experienced good success with obtaining research funding since the new directions were high-priority areas for US government funding agencies, as well as industrial research organizations. My area of research is far from unique in this regard. The main point is that you need to carefully consider the future prospects for your research area, and whether the area in which you perform research is maturing, or if new and novel approaches are emerging or have the potential to emerge. While it's not possible to predict all future events, some trends make themselves known at a fairly early stage of any given technology or scientific area. To accurately assess the future prospects for your research, you need to stay current in your knowledge and monitor the developments reported by your professional colleagues and competitors. Similar situations exist for many other research areas, and it's very important to stay abreast of scientific and engineering developments and to attempt to keep your research at the state-of-the-art. To not do so will quickly make your research obsolete, and you'll find it increasingly difficult to obtain research funding as your technology area matures, while moving to the new and emerging areas can position you for future success and sustainability.

It's very important to recognize that learning is a lifelong process and that you need to spend significant time and effort upon keeping your knowledge current. This not only helps you to be an effective researcher, but it helps you recognize when research topics mature and similar areas emerge. This recognition is a key to continuing success, and the first researchers to work in a new area always have the easiest path to becoming established. The good news is that it's often not difficult to transition your research efforts to new areas, while maintaining the majority of your current knowledge base. The transition to a new or emerging area is often a fairly simple extension of your current knowledge base, and all that's required is to learn about the new area and tailor and adapt your current knowledge to the new research topic. Some study is required, and you need to read the published literature in the new area, and attend some conferences or workshops to learn the current theories and practices for the new area. You will also learn the identities of the research performers, which is a good mechanism to start to build collaborations. You should attempt to communicate with these researchers and initiate discussions that can help build relationships. Finally, there is definite benefit to being one of the first researchers in a new or emerging area regarding the ability to obtain research program funding.

Funding agents and US government program managers also follow the latest developments and generally have interests in new ideas that can build new and emerging areas in their interest areas. They routinely participate in strategic planning workshops and meetings, both within their own agencies, within government-wide strategic planning activities, and internationally. Emerging scientific and technical areas are identified and there is high interest in supporting the latest developments in which researchers and organizations are establishing the state-of-the-art. This process provides the opportunity for new research support possibilities.

The first researchers working in these areas generally have a much improved success rate in obtaining funding, compared to those who follow later. Working in a new and emerging area also presents the opportunity to become more easily established since, in general, you're not necessarily competing with well-known researchers in a mature area. While established researchers will, undoubtedly, also seek research funding to work in these areas, their ideas will not, necessarily, be more attractive to program managers than your own. The program managers are trained to look for new and young researchers with novel ideas and approaches to problems. This provides a unique opportunity for recent graduates and young researchers to obtain research support funding. You should always keep your knowledge current and be aware of these new and emerging research areas. Science and engineering areas continually advance and evolve, and new and novel directions and concepts are continually being proposed and developed. Success in pursuing new areas and directions generally involves making a connection between existing and established concepts, and exploring how the established science and technology relate and lead to new advances. You'll find yourself engaged in this process, which will likely dominate your thinking and actions throughout your career. The willingness and capacity to adapt to a changing environment, and to pursue new and emerging areas offers excellent avenues to further your research and to maintain your research career throughout your life.

4.9 Self-Evaluation and Projecting Confidence

In the process of deciding what research area you plan to pursue, there are two major factors you need to consider: (1) in what scientific and technical area do you have the most knowledge and feel most comfortable in pursuing, and (2) what research areas and topics offer the best opportunities for obtaining research funding? What you are best at, and what topics you wish to pursue clearly need to be major considerations. You are most likely to make the greatest progress and have the most success working in an area that you really enjoy. To be truly successful, your occupation and research endeavors need to become your hobby. However, you should also consider where you are likely to have the best chance to compete for research funding. Research funding opportunities can either be offered on a periodic and regular schedule, or they can be directed one-time opportunities. You need to stay aware of when scheduled research grant solicitations arise, and regularly monitor funding agency announcements. Many universities will monitor these grant opportunities through their sponsored research office and distribute the information to their faculty members. If your university offers such a service, you should make sure you are on the distribution list. Otherwise, you'll need to monitor this information yourself. Regularly surveying funding agency websites offers an easy method to complete this function. No matter the particular research announcement schedule, the grant funding solicitations all have announced end dates for proposal submission that need to be honored. This means that you may need to temporarily cease to work on other projects in order to prepare and submit a proposal by the due date.

Once you identify the research program opportunities you intend to pursue, you need to project confidence. When you contact and communicate with program managers and program directors, you need to present yourself as a very knowledgeable and personable researcher with a thorough understanding of the scientific and technical issues associated with your research. You need to know the identities of other researchers working in the field, and what major discoveries and developments have been reported. You should know the identities of other researchers pursuing work similar to your own and, in fact, researchers who could be possible collaborators. We like to think that we're evaluated solely on our research ideas and performance, but our personalities and how we interact with our colleagues and other professionals are also major considerations defining our success in building research programs. Personal traits are important factors for successful interactions with all program managers and program directors since they need to have confidence in you, along with your research proposal. You need to convince them that you really know your subject and area, that your proposal is at the state-of-the-art, and that you are ready to make significant advances in your research and your chosen field. While a well-written proposal clearly is very important in establishing this image, personal interactions with the program manager or program director can have a very significant and positive effect, and can make the difference between receiving good proposal ratings, and actually obtaining research funding. These two issues are not necessarily the same.

Keep in mind that in the highly competitive research-funding environment that currently exists, many proposals receive good and sometimes excellent ratings, and are rated sufficiently high that they could receive funding, but they aren't selected. The government program managers and program directors generally have several highly rated proposals in any given competition, but available funding for only a small fraction of the proposals. Authors of the highly rated, but not selected, proposals are understandably disappointed when their proposal is not selected for funding, and sometimes they are extremely upset. They will attempt to contact the program manager or program director, or even the division director, and argue for their proposal. I've seen situations as a program manager and division director where the Principal Investigator (PI) of a declined proposal that had received good or excellent ratings was extremely irritated. However, once the decision to not fund a proposal has been made, the chance of anyone reversing the decision is remote, at best. For mission agencies there is generally no or limited opportunity to challenge a review. The NSF does have a formal procedure for principal investigators to challenge a funding decision, but a reversal of the decision not to fund is extremely unlikely. The NSF's procedure involves formation of a committee within the directorate, and outside the particular division to which the proposal was submitted. Generally, the committee will be formed in another division within the directorate. The committee will review the declined proposal, directing their attention to the reviews that were performed, along with the proposal ratings, and will consider the decision that resulted. However, the evaluation is primarily focused upon the proposal evaluation procedure and whether or not the committee feels the proposal evaluation was conducted in a fair and unbiased manner. If the committee concludes that the review and evaluation were fairly conducted, they will support the original review results. Scientific or technical details included in the proposal are rarely addressed. Also, the committee will not address issues associated with a proposal being declined for lack of available funds, which is a division or directorate budget consideration. The lack of available funds is the most common basis for declination of highly rated proposals. Therefore, the success in obtaining a reversal of a declined proposal decision is unlikely, and generally, not worth the effort of the PI.

The major question becomes: How do the program managers or program directors make the decision as to which proposal to select for funding? The answer can be complex, depending upon a variety of factors, ranging from the subject and topic of the proposal, to the issue of agency or national research priorities, to issues related to the PI. They will generally rank the highly rated proposals based upon consistency with division, directorate, and agency priorities. However, their personal knowledge of the PI is important. For example, how the program manager or program directors rank or order the proposals in the highly rated proposal group often depends upon their personal knowledge of the PI and his or her previous research and reputation. They need to have confidence that they are selecting a proposal from a PI likely to make good progress in the research. This is one of the reasons you want to personally meet the program manager and have the opportunity to discuss your research. In these meetings you don't want to be shy, but rather to exude confidence. You want to be personable and knowledgeable. Your overall concern is to give the program manager or program director confidence in you as a researcher and to assure them that you and your research are worth their agency's financial support.

4.10 Group Research Activities versus Single Principal Investigators

As you consider your research area, you might also want to consider how you will pursue your research goals and how you will go about building your research program. You should consider if you wish to pursue your research in a group environment or as a single investigator. How you wish to proceed is dependent upon your personality and how you best function. There are advantages in either approach, and success is possible from both modes of operation. In my experience, successful faculty members all seem to have a few common characteristics and traits. First, they are all competent and have good knowledge of their chosen field. They are all experts in their scientific and technical disciplines and have a developing or proven track record regarding their previous research and associated professional publications and conference presentations. Second, they are all motivated self-starters with a strong desire to succeed. Some of them, but not all, are intensely competitive and believe their research is on a par with anyone, anywhere. These competitive faculty members generally end up being the leaders in their departments and research groups, and aggressively compete for research grant funding, generally responding to each and every grant opportunity that is appropriate for their desired research area. These are the people who would visit me on a routine basis while I served as a program manager for the DOD, and as a division director for the NSF. These people were all very focused upon their goals, seeking every opportunity to make contact with program managers and program directors and learn all they could regarding paths to success. Once a grant funding opportunity was identified, these researchers would take the lead to get a proposal organized, prepared, and submitted. Often, they would recruit other faculty members and researchers to participate in the work, organizing focused programs to address the proposal opportunity goals and requirements. My observation is that the less-competitive faculty members, while very competent and skilled, would tend to work in group activities and would participate in the actual performance of the research projects, but would not be the leaders.

Research groups generally self-assemble with someone taking on the role as leader, who takes responsibility for making contact with the funding agency program managers and program directors, while other members of the group serve as participants and contributors who make very significant contributions to the actual performance of the research, but do not take responsibility for the initiative to organize and submit proposals. Both types of researchers will mentor and advise graduate students, serve as joint authors of papers and technical presentations, and participate in research program reviews. These group activities tend to be very effective and productive, and participation in a research group is a very attractive and effective way to become established, as group research tends to progress more rapidly than the single researcher working on his or her own. Research groups tend to support fairly large numbers of graduate students, who collaborate and work on projects that support each other, or projects that are components of a larger research project. However, not all faculty members desire or wish to work in a group environment. Some faculty researchers respond best when working on their own with their students, which is normally a small number. You need to consider your interests in group activity, or if you prefer to work on your own. If you are interested in participating in a research group, you can explore joining an existing group, or you can assemble your own.

4.11 Success-Breeds-Success

Of course, as stated throughout this book, you'll need to acquire funding to perform your research (the main subject of this book), which will consist primarily of research grant funding, but will also derive from fellowship and scholarship support for your students, industrial support, both financial and in-kind, and other grant and financial support opportunities that will periodically arise. You'll find your ability to raise and obtain research funds becomes easier as your research output increases and you become more recognized as an active contributor to the research area in which you work. You'll find that a direct feedback-effect relationship between research output and the ability to sustain a well-funded research program exists. The basic principle in research program development is "success-breeds-success." That is, the more quality research you contribute, the easier it becomes to write successful proposals and to obtain funding for your research activities.

The research community in any given technical or scientific area is actually a fairly small community, and essentially all the major active researchers and grant funding program managers know or are aware of each other, personally, professionally, or through the published literature. They attend the same conferences and meetings where they interact and communicate and exchange ideas, thoughts, theories, and the latest scientific results and directions. They are well aware of emerging ideas and concepts, along with what progress is being achieved, and the research topics that are not showing or demonstrating progress. They also read each other's publications and are well aware of the research directions that are being explored and pursued by their peers, collaborators, and competitors. This information helps them define their own research directions and approaches. Funding agency program managers and program directors participate in these meetings, and generally are well aware of the major directions and approaches pursued by the major research contributors, not only nationally, but also internationally. In fact, knowledge of the state-of-the-art in specific technologies and the underlying science and theoretical foundation is a major occupation requirement and expectation of government funding agency program managers and program directors.

4.11.1 Funding Agency Program Director Visits

Once you make contact with a program manager or program director from a grant funding agency you want them to visit you. As a US government research program director and manager I often visited individual researchers, toured their laboratories, and discussed their research approaches and results. These visits are often used by funding agency program managers not only for the purpose of reviewing the progress on funded grants, but also for learning of new approaches to research problems and recent developments. This information is subsequently used as input to evaluate progress in the field and to evaluate progress made by researchers that they are supporting, and also to define new research directions and grant funding opportunities. Program managers, particularly from mission agencies such as the DOD, DOE, etc., generally visit a fairly large number of researchers working in scientific or technical areas in which they have interests. Some of these visits will be for the purpose of being briefed on the progress of research they are funding, some visits will be in response to invitations to give talks or tour facilities, and some will be for the purpose of visiting a researcher that has been identified as performing interesting research with a novel approach.

During these visits the program managers are generally well aware of the research being discussed, the identity of other researchers performing similar research, and the state-of-the-art of the research being reviewed and how it relates to current research efforts in progress elsewhere. The questions the program manager asks are a good indication of the depth and maturity of their knowledge of a specific research topic. For example, the program manager will often ask very probing questions, but will not actually provide or reveal a good description of their detailed knowledge. They are trained to not reveal information they learned in confidence from another scientist or engineer, who in fact, may be a direct competitor to your work. They will never reveal information they learned from reading the many proposals that they receive, or they learned from discussions with other researchers who presented their results to them in the process of applying for a research grant. Program managers realize that all researchers seeking potential funding for their work will disclose confidential and proprietary information that is not to be disclosed to others. The program managers will only mention the work of others if it has been openly presented in publications and reports, or at meetings open to the professional and scientific communities. However, during the visit there is generally a very open, frank, and comprehensive discussion concerning the research that is being reviewed. The program managers are not only interested in learning about the research progress, but they are also interested in meeting and learning about new and young researchers participating in the research, particularly those in the process of establishing their careers. Truly creative and innovative research, and the young engineers and scientists performing the research, are identified and this identification facilitates both new grant opportunities, as well as successful follow-on grant proposal awards for those identified as state-of-the-art researchers.

Once a funding agency program manager identifies a new researcher who is performing creative and innovative research in an area in which the program manager or their agency has interest, that researcher has made a major advance on their way to becoming established, and the researcher's ability to obtain a research grant and to sustain a funded research program is enhanced. The program manager may become an advocate of your work, particularly if your research supports their research interests and goals, and they may be willing to help you to secure grant funding, either from their organization, or from other organizations that may be more appropriate for the research topic. This is the essence of the "success-breeds-success" principle. The process starts with the performance of quality and innovative research, which provides the basis to build and enhance your program, and it is for this reason that you want to get as much exposure for your research as possible. You want to expose your research results and progress to as many research grant funding program managers and program directors as possible. You want to be on their "radar" and to make them aware of your work. You want to visit them and discuss your research, as well as have them visit you. Many faculty members, particularly when they are attempting to become established, will make regular visits to funding agencies and attempt to discuss their research with as many funding agency program managers as possible. For example, as a US government program manager I worked with one faculty researcher who would visit me on a regular basis. In fact, I noted that he would make an appointment for his next visit with my secretary upon leaving my office after virtually each meeting. This went on for a couple of years. The tactic in the end was successful, as the faculty member was able to identify a subject of mutual interest that related to his research, as well as the agency interests, and he successfully obtained funding.

4.11.2 Maintaining Communication with Program Managers and Program Directors

Many faculty members also maintain a list of funding agency program managers and routinely send them copies of papers that they publish. As a US government research program manager I've been the recipient of many copies of publications that have been mailed to me, either in regular mail, or by email. Typically, I would scan the papers looking for topics in which I had interest, and I would read those papers. I would also scan the other papers in which I had some familiarity and sometimes, but not always, read those papers. I would generally not read the papers in which I had no interest, unless they were on a topic which I had heard about, but of which I had little personal knowledge. In that case I would read the paper as an introduction to the subject for me. This was particularly helpful to me if the subject was one that was considered as being of emerging importance. In these cases, having the paper sent to me was actually a service that I appreciated, and I would remember the author's name. For the majority of the papers, the author's name and the subject of the paper would generally stay in my memory, and later be recalled, particularly when reading similar work, either reported by the same author, or by other researchers working on similar topics. If the research topic was timely, and of significant current interest, I would be interested in reading essentially all the papers on the subject that were sent to me. Follow-up face-to-face meetings occasionally resulted, which presented the opportunity for more detailed discussions. This type of information has a way of circulating in a program manager's mind, and when the program managers read research ideas, either in white papers or submitted proposals, the information helps them place the new research idea in perspective, as well as make the program manager aware of the identity of those in that research community. The lesson here is that you want the program managers to identify you as a contributor in your research specialty, and any method or action that results in exposure of your name and your association with the technical subject is a positive development.

Exposure to your research can initiate a process that makes the program manager aware of what you are doing and trying to accomplish, and if the work is in an area in which they have interests, the program manager will start to recognize you as a promising researcher with new and creative approaches and ideas. This can, of course, lead to the submission of a proposal and award of a research grant. To facilitate this process, you need to publicize your research approaches and research results to the greatest extent possible, and you want to publish in professional journals, present your results at conferences and professional meetings, and accept as many invitations to present your work as possible, whether at a specific company or small meeting, or at large and major conferences. When you have a paper published you should also consider sending it to appropriate research funding agency program managers. In fact, for the reasons stated, I recommend sending copies of your published papers to program managers for their use. You may not receive an acknowledgement or any feedback, but you may help the program manager become aware of your name and the type of research you are performing. This will facilitate future communications with them. Program managers are more likely to respond to communications from researchers who they either know personally or at least know by reputation. Also, the fact that the research results have been published is an indication that the research is high quality and original. Although the submission of your research publications to them is unsolicited, as indicated, the publication can serve as an introduction to the program manager of you and your research. It follows that the "publish or perish" principle is true, and publication of your research needs to be an extremely high-priority goal for you.

Publication of research results is a very important factor in building a funded research program, and there is essentially no downside to publication and professional exposure through presentation of your work at conferences and professional meetings. This activity is a main fuel required to drive the engine of a successful academic research career. You need to make publication and presentation of your research a prime priority and to develop the habit of submitting your work for review and publication. Of course, you want to publish your work in the highest-rated journals in your field, and present your work at the main conferences and meetings that are devoted to your technical field. Every scientific and technical field supports a main or limited number of primary publications and conferences. Publication and presentation of your research results in these forums should be your highest and most immediate goal. However, often specialized publications, conferences, and meetings are also organized, particularly for topics of significant current interest. These focused-issue forums and venues are also high-priority opportunities for presentation of your work since they are well attended by the main community pursuing research in that area. The attendees at these meetings generally consist of active researchers and funding agency program managers who are directly interested and involved in the research approaches and results that are presented on the topic.

4.12 What We've Learned

In this chapter we discussed how a new faculty member goes about getting started in building his or her research program. We assumed you had just accepted a faculty position in a research-oriented college or university, and that you were in the early stages of deciding your research area and topic. The point is emphasized that you are in a similar situation to someone who is starting a new business, and we discussed the entrepreneurial dimensions of an academic research career. You need to define your products, customers, marketing procedures, and revenue and resource sources. The point is made that, just as for a start-up business, the majority of failures occur for the same reason; lack of focus. To avoid this, the need to perform self-evaluation of your personality, current knowledge, and your preferred research area must be completed. You also need to determine if you plan to build your research program on your own, or if you prefer to work within a group, which can be an existing group, or one you plan to organize. In either event, the group activity will involve multiple faculty members. The need to publish your research results, and to build relationships with grant funding agency program managers and program directors, and techniques to accomplish this were discussed.

5   Know Where the Money Is

For the vast majority of academic researchers there are two primary and major sources of research funding: federal government agencies and industry. These sources, in fact, provide, on a regular and continuing basis, the main funding available for performing research in academic institutions. The federal agencies, in particular, have a long and distinguished record of providing funding for academic research, and have very well-developed policies regarding the amount of funding they provide, including a definition of the performance period, which is often multi-year, as well as the policies that govern how the funds are provided and managed. Industrial and business sources can also be a significant research-funding provider, although these sources generally provide a lower level of funding, and often dictate shorter performance periods. However, there are additional sources of research funding: state and local government agencies, not-for-profit foundations, and other private sources also provide funding for academic research and should not be neglected in the effort to build one's research program. Also, new approaches to obtaining research funding periodically emerge, including a very interesting "new" approach to securing research funding that has recently developed and is termed "crowdsourcing," which is a technique that makes use of the internet and social media to raise money for specific topics. This unorthodox technique, which has so far been used for only limited amounts of academic research funding, has nevertheless started to be used by some faculty members to support their research. This method, at this time, is too new to have any significant history, but is an area that you might want to monitor as it develops.

In this chapter we'll address the various sources you can investigate for opportunities to fund your research. If you're a new faculty member, recently hired, you probably don't have a good idea about what agency you should approach, or who in that agency to contact regarding funding for your research. You may know of the government funding agencies and may have even tried to contact them, most likely with limited success. Your first challenge will be to identify the agencies and offices that support the research that you are interested in pursuing, and then try to identify the appropriate program manager or program director who might be interested in funding your research. You'll also need to learn how to approach them, make contact, and establish effective communication, which we'll discuss in the next chapter. You'll want to learn how they function regarding proposal submission interests and requirements. Identifying the appropriate funding agency and how you can interact with them can be a daunting task, until you learn the basic principles. In this chapter we'll discuss the fundamental principles associated with working with research funding agencies. We also present a brief discussion of research funding trends over time, and the various methods federal agencies use to advertise their research interests and funding opportunities.

5.1 Basic Principles for US Government Research Funding

Before proceeding to learn the basics of how to identify the appropriate funding agency or office for your research topic, and the appropriate research manager or program director with whom you need to communicate, it's helpful to learn a few basic principles and facts concerning US government research funding and how it operates. These basic principles apply to all funding agencies and sources. In this regard, I'll address five fundamental principles that you should understand.

Principle One: The eligibility of scientists, engineers, and researchers to submit research proposals to US government agencies is defined, based upon specified criteria.

The US government offers research support funding to all qualified bidders. This means that anyone who satisfies certain and defined basic criteria is qualified to submit a proposal, and that their proposal will be fairly evaluated and considered. For academic researchers, to satisfy the basic criteria generally means that you hold a tenure-track or permanent research-qualified faculty position in a US college or university in good standing with the government. If so, you are eligible to submit proposals seeking financial support for your research. All faculty members that are on a tenure-track position or are tenured are eligible to submit research proposals to any US government research funding agency, and this privilege extends to all private and industrial funding opportunities, as well. Research-qualified can also mean that you are in a non-tenure-track position and have a title such as Research Assistant Professor, Research Associate Professor, Research Professor, Researcher, Scientist, or other title that indicates permanent employment with duties that are directed towards engagement in research, etc. In general, you need to be a faculty-rank researcher employed by the university. Generally, those with titles such as Post-Doctoral Research Assistant (generally called "Post-Docs"), Visiting Professor, Visiting Scientist, Adjunct Professor, or other title that implies a temporary, time-limited position, or a non-permanent position, are not normally permitted to submit research proposals on their own. However, most of these research personnel can be named on a proposal as an Investigator, and sometimes as a Co-Principal Investigator, although they generally cannot be listed as the Principal Investigator (PI).

Good standing with the government means that neither the PI nor the institution have been found to be in violation of official government policies or regulations relative to their research activities and financial management procedures. If a researcher has been convicted of an ethics or performance violation they may be barred from proposal submission. If an institution has been found to be in non-compliance with US government policy or laws, generally discovered during the performance of an audit, the government may disqualify researchers from that institution from submitting research proposals for a period of time, generally until the institution takes appropriate corrective actions, in accordance with appropriate policies and law. If this were to occur, your institution will notify the research faculty members of the situation and will not approve the submission of research proposals, in accordance with their status.

Principle Two: There are several different categories of research funds, defined by the maturity and sophistication of the research.

The research funds offered by the government, particularly those offered from mission agencies such as the Department of Defense (DOD), the Department of Energy (DOE), NASA, etc., come in different "categories," consisting of "basic research," "applied research," and "advanced development." The DOD, which is a major source of research funding, terms the basic research, applied research, and advanced development categories as 6.1 basic research, 6.2 applied research, and 6.3 advanced development, respectively, following the categories that are listed in the federal budget approved by Congress for the defense department. The 6.1 basic research category historically has been used to support research that is general in nature, and intended to discover fundamental laws and principles. There usually are no direct applications or systems use required, although the basic research will establish the foundations for more advanced and applied applications. Research projects funded from this category generally do not include "classified" research projects performed or supported by the defense and intelligence agencies. There are cases where certain basic research projects may be "classified," but when this occurs the classified nature of the research is stated in the Call for Proposals. This topic is discussed in more detail in Chapter 8.

The results of basic research projects are almost always widely reported and published in the open literature. In fact, publication in the open literature and presentation of research results from basic research funding is generally an expected and highly desirable outcome of the research project. Publication of research results is often used by program managers as an indication of the quality of the research, and provides evidence that progress is being achieved and demonstrated. A significant fraction of the research funds provided by mission agencies for university research come from the 6.1 basic research category. Also, the vast amount of research support funds provided by agencies such as the National Science Foundation (NSF) and National Institute of Health (NIH), consist of funds in the basic research category.

The 6.2 applied research category provides funds to pursue projects designed to advance a given technology that already exists or has been researched, identified, and demonstrated on basic research projects. Often these projects focus more upon the development of a new device or system, with the goal of improving the performance, manufacturability, or cost of a given technology. The overall plan is basically a linear model where a new concept is defined and developed with basic research (6.1) support and is then advanced to a more practical realization under applied research (6.2) funding. The advanced development (6.3) category is intended to further develop and advance the technology towards practical use, insertion into systems, and commercial realization. Most of the funds available in the advanced development (6.3) category are provided to industrial and government laboratories. Academic institutions and researchers are not generally major recipients of funds from this category, although some of these funds do go to academic researchers, particularly when they work on collaborative research projects with industrial or government partners. The amount of funds provided in the three categories increase as ideas and concepts mature and develop, with the basic research category receiving the lowest amount of funding, and the 6.3 advanced development category receiving the largest amount of funding. Generally, the research funds provided to academic researchers come primarily from the 6.1 and 6.2, basic and applied research, categories, although, as mentioned, some 6.3 advanced development funds are sometimes provided, based upon the particular technology and project. The 6.1 and 6.2 basic and applied research categories are combined and called the Science and Technology (S&T) budget in the US Department of Defense budget. For academic researchers the S&T budget is the one of most interest since this budget will provide the vast majority of funds available to support their research activities.

Principle Three: Research funding agencies do not all have the same goals or reasons for providing research support to non-government performers.

Certain US government funding agencies are considered as "mission" agencies and some are not. The "mission" classification indicates that the agency exists to perform a well-defined purpose with a stated mission. For example, the DOD exists to provide for the nation's defense and security. The NIH exists to provide for the health and welfare of the nation's population, and to advance medical science and technology that will serve all the world's people. Likewise, the other major departments that support external research do so in the context of their overall mission. It is in this capacity that these departments, through their various agencies and offices, provide research funds, and all research projects that they support somehow fit into the overall mission of the agency. The exact research topics that each agency will support are determined by the agency through their internal strategic planning process. This is an important factor you need to consider in building your research program since you'll most likely need to modify and "tune" your proposal to areas of interest to the agency to which you submit your proposal. Your goal will be to determine what these interests are and decide how your particular research can fit into the program appropriate for your research.

The NSF is an example of a non-mission agency. That is, the NSF does not have a goal-oriented mission related to a particular product or service, and does not exist, in general, to support specific and particular scientific or educational projects of direct use to the government, or engineering technologies for stated and well-defined purposes. The NSF was established to support science, engineering, and education research and development activities, in general, based upon quality metrics. The NSF is essentially in partnership with the academic community and generally makes their decisions on what proposals to fund based upon a peer-review process, conducted by scientist and engineer colleagues, often organized into proposal review panels. Decisions to select a proposal for funding are determined by proposal evaluations based upon the basic quality of the proposed research as a primary metric, and the principal investigator submitting the proposal determines the subject of the proposal. The quality of both the research being proposed and the preparation and ability of the principal investigator are the main criteria for making funding decisions.

For "open-window" NSF solicitations no restriction is placed upon the research topic, other than the subject of the research topic be in the discipline area of the division to which the proposal is submitted. Also, the NSF offers research support under both their "core" program, as well as for specific, well-defined research programs. The "core" programs have time-limited "windows" open once or twice each year, depending upon the division or directorate, for submission of unsolicited research proposals. These windows will accept research proposals from qualified institutions on virtually any research subject, with the caveat that the research subject must be appropriate for the particular division or directorate to which the proposal is submitted. However, this is the only restriction on the research subject matter. The other research solicitations are focused upon select, well-defined research topics, which are described in the program announcement (i.e., the proposal solicitation). The NSF often publishes what is called a "Dear Colleague Letter" to announce these research opportunities. These special research proposal solicitations are publicly announced, as they are defined and developed, and you need to stay alert to be aware of these opportunities since they generally aren't announced on a specific pre-determined calendar schedule. However, all program solicitations are well publicized and are available on the NSF website, as well as the Grants.gov website, where all details related to the particular opportunity are indicated and defined.

The mission agency designation affects the manner in which the government provides research funding to outside organizations. The government considers research, particularly that solicited by mission agencies, to be essentially the same as a service or product it acquires from outside sources. In this regard, research is something that the government "buys" from an outside vendor. As such, the research topic or research funding opportunity is generally defined by US government agency program managers, and the research is desired and intended for specific and particular purposes or applications in order to advance a specific technology, application, or system. The research topic details, which may be listed in the form of product specifications, are always well defined in the Call for Proposals. Unsolicited proposals that do not address the topics of interest to a particular agency will not be accepted.

Since research is treated as a product or service that is being purchased, mission agencies expect to pay the full cost of the research and generally, but not always, do not require the proposing institution to provide any "cost-sharing" or "matching" funds, where any of the cost of the research is borne by the proposing institution. If cost-sharing is required, the Call for Proposals will clearly state the terms and conditions of the requirement, along with the applicable details.

There are two types of cost-sharing requirements: mandatory and voluntary. Mandatory cost-sharing requirements are applied to certain research solicitations. When this occurs, the requirements are clearly identified and stated in the research solicitation, and the proposal budget must clearly indicate that the proposing institution is providing these funds. For basic research (i.e., 6.1 research solicitations) mandatory cost-sharing is not common. This is not the situation for non-mission agencies, such as the NSF, which works more in partnership with academic research institutions than as a purchaser. That is, the NSF exists to assist in the pursuit of fundamental scientific, engineering, and education research, and to assist academic researchers in their research endeavors. As a partner in this effort, the NSF expects the host institution to provide some of the required support for research projects. The NSF considers research to be part of an academic faculty member's job responsibility and, therefore, it will not generally provide support funds for a portion of the faculty member's academic year salary (generally called "release time" or "academic year" salary). In this regard, the academic institution is essentially supplementing the research project by supporting the principal investigator's time while they work on performing the research. This requirement is essentially limited to the NSF, and other US government agencies have no limitation on providing a portion of the academic year salary for a PI or other faculty researcher.

The situation is different for the summer months when most faculty members do not receive salary from their institutions. All government research-funding agencies permit the payment of summer salary for researchers who perform research during the summer months. The NSF, however, does have a limitation on the time that can be committed to summer research, and there is a one-month summer support limitation for a PI working on a given project, with an overall limitation of two-months summer support that can be provided for all NSF supported projects. Mission agencies, such as the Department of Defense or the Department of Energy, do not have a limitation on the length of time during the summer that a researcher can be employed performing research, either on a single project or in aggregate, except that a researcher cannot commit more than 100% of the summer period to the total of his or her activities. Voluntary cost-sharing commitments are sometimes offered by proposing institutions, generally with the hope that such a commitment will enhance the prospect that the research proposal will be accepted and funded. These voluntary cost-sharing offers and commitments are not required, and the NSF explicitly prohibits voluntary cost-sharing for unsolicited proposals. Government program managers, or external proposal reviewers, are not permitted to consider voluntary cost-sharing in their evaluation of research proposals for basic research. Therefore, voluntary offers of cost-sharing are not necessary or helpful. When mandatory cost-sharing requirements exist, they are always clearly stated.

Principle Four: All research-funding opportunities offered by the US government must be publicly announced and advertised.

The question for a new researcher becomes: Where are the research funding opportunities announced and published, and how does one start to search for research funding opportunities appropriate for his or her specific research topic? More importantly, how does the researcher find out which funding agency is interested in their specific research topic and to which agency or office do they submit their proposal? These questions are, of course, very important and, quite frankly, the main topics of this book. We'll discuss these questions in more detail in the following section of this chapter. However, to start, we recognize that the basic principle is that the government must permit all qualified bidders to submit proposals for research projects that the government wishes to support (as discussed in *Principle One)*. Therefore, the US government research funding agencies need to advertise the topic areas in which they are interested in accepting proposals, the details on where to submit proposals, and the guidelines for proposal submission. This is accomplished using the following methods.

5.1.1 Broad Agency Announcement (BAA)

A main mechanism to announce opportunities for research funding to the public is through a process that makes use of a Broad Agency Announcement (BAA). This mechanism has evolved into its present format over time, and the BAA is now extensively used to advertise US government interests in supporting research for specific areas and topics. The government has supported certain non-government scientists and engineers working in research areas and topics of interest to the government dating back to the founding of the country. However, in the early days the process was not extensive, well defined, or consistently applied across government agencies. As discussed in Chapter 2, widespread government support for academic research only became common after the Second World War. Many of the research projects were funded on a specialized case-by-case basis, and some were not widely published or open to everyone that might be interested in competing for the research funds. There was a need to establish a standard and common mechanism to publicize the areas in which various government agencies had research interests, and to make their research funding opportunities known. The BAA solicitation process evolved, and it became more formal when it was redefined when Congress passed the Competition in Contracting Act of 1984 (Public Law 98–369). In this act, Broad Agency Announcements (BAAs) were created as a method for government agencies and organizations to define research areas and topics in which they have interests in supporting research and to solicit proposals. All BAAs are published and available to anyone eligible and interested in competing for the research funds.

The BAAs describe basic (6.1) research and applied (6.2) research topics and do not include demonstration projects or validation of previously performed research. The main intent of the research announced in a BAA is to increase scientific or engineering knowledge in a given area and, as such, multiple and varying approaches are generally supported. The BAA basically outlines the research interests of a government agency, and it is a solicitation for interested and qualified parties to submit proposals to perform research defined in the BAA. The BAAs are published periodically and are used both to describe the range of research topics of interest to the issuing agency in general, and/or to describe specific funding opportunities for research topics that are of interest to that agency at a particular time. In the former case the BAA may be open for proposal submission on a long-time basis, typically a year or more, and in the latter case, the BAA may be open for proposal submission on a short-time basis, typically one to three months. The amount of government funding available for research described in the BAA may be limited, particularly for the more general long-term BAAs. In some cases, the issuance of a BAA does not indicate or guarantee that any funds are actually available, and only indicates the research areas of interest to the particular agency. Publication of the BAA permits the agency to fund projects defined in a proposal at any point when funds become available. This often occurs in response to unsolicited proposals that are submitted, and there is interest in the funding agency. In this case, the publication of the BAA simplifies the procurement process.

Conversely, a BAA issued for a specific topic or research-funding opportunity will generally indicate an immediate funding opportunity, and the BAA will state the amount of government funding that is available for the particular project described in the BAA. The BAA document will include a detailed description of the research topic, as well as the criteria associated with the award, the proposal requirements (i.e., what needs to be included in the proposal), and a description of the proposal evaluation criteria. Often the BAAs will contain a detailed list of requirements for the project, as well as detailed proposal format and submission details. It has become common in recent years for proposals to be limited in size, that is, limited in the number of pages, and this limitation will be indicated in the BAA. It's very important to carefully read all proposal requirements, and to respond with a proposal that is in compliance with all requirements. Also, the BAA may indicate the necessity to submit a "white paper" before formal proposal submission. This issue will be described in more detail in Chapter 6.

When responding to a research topic or area announced in a BAA, it's always best to contact the appropriate program manager or agency contact person, generally indicated in the BAA, before preparing and submitting a proposal. The agency contact can provide additional information and provide an indication if your idea and research approach would be appropriate for that particular opportunity. The contact information, including email address and phone number, are generally available either in the BAA, or on the agency website. Sometimes only the email address will be available. Please keep in mind that the program managers and program directors are busy people, and that competition for funding is intense, and therefore, they may wish to be contacted by email, rather than by phone. However, it is their duty to respond to inquiries, so you should get a reply, although there may be a delay.

5.1.2 Other Research Opportunity Announcement Mechanisms

A variety of other mechanisms, similar to the BAA, to advertise funding opportunities are also used. The National Institutes of Health (NIH) advertise their research grant opportunities as Funding Opportunity Announcements (FOAs), which can be published as either a Program Announcement (PA), or as a Request for Applications (RFA). The NIH Institutes and Centers will accept unsolicited proposals for research topics that do not fit well within one of their FOAs. The National Science Foundation (NSF) publishes their focused research grant opportunities as a Solicitation or a Call for Proposals (CFP). The NSF will often use an announcement procedure called a Dear Colleague Letter (DCL), which describes the research funding opportunity. All funding opportunities will be published on the NSF website, Grants.gov, and also announced to the academic community. Unsolicited proposals can be submitted to the NSF during the fall and spring open window periods. The dates for the open window periods vary slightly from division to division, and from directorate to directorate, but they are all advertised on the websites for the various divisions and directorates. The Department of Defense (DOD) and the various DOD offices and agencies advertise their specific research grant opportunities through the BAA process, and through Requests for Proposals (RFPs) that are issued for specific research opportunities. Unsolicited proposals can be submitted pretty much any time of the year, although it is always best to communicate with a DOD program manager before preparing and submitting an unsolicited proposal, both to determine their interests in your work, and to determine if funding is available. Each research office and research agency within DOD publishes their own BAA, appropriate for their specific research interests. The Department of Energy (DOE) advertises their research grant opportunities by means of Funding Opportunity Announcements (FOAs). All proposals submitted to DOE must be submitted through the Grants.gov website (described in Section 5.2.2). The DOE Office of Science does not accept unsolicited proposals. NASA advertises their research grant opportunities by means of NASA Research Announcements (NRAs). More details on each of these agencies are presented in the following chapter.

Principle Five: Several legal mechanisms (instruments) are available to transfer research funds from US government agencies to nongovernment organizations.

The mechanism by which the US government provides support funds to research performing organizations is defined by agreement between the US government and the academic community, and the US government has several legal instruments available to them to fund external research and/or to secure scientific and engineering services from nongovernment sources. These instruments provide the legal basis for providing US government funds to external organizations to secure research that the government wishes to pursue and support. These legal instruments consist of (1) contracts, (2) grants, (3) cooperative agreements, and (4) a category called "other transactions." All of these instruments are used, depending upon the agency, and the specific program or funding opportunity. The four instruments are briefly defined as follows.

5.1.3 Contracts

A contract is a legal instrument negotiated and signed between two parties, stating the agreement between them, including definition of deliverables, performance period, and financial requirements, to acquire certain products or services. The US government makes extensive use of contracts to acquire a wide and diverse array of products and services, and research is often secured by means of contracts. From this perspective, university research is simply a service contracted for, and provided to the government. A contract always requires some sort of "deliverable," which can be a product or service. At times progress and final reports may be all that is required, while at other times prototype or demonstration hardware or software may be required. The actual and specific deliverables will be carefully defined and described in the contract, which is executed and signed by both parties.

Research performed in industrial laboratories for the US government is almost always secured by means of a contract, which states an agreement between the company and the government to provide a range of services, which often include deliverables of working materials, devices, circuits, systems, etc. It is not unusual for academic scientists and engineers to work with contracted companies or business organizations to assist in the work. Quite often the company will employ what is called a "flow through" arrangement with the university, where the same legal requirements that exist in the master contract are applied to the university. This can be burdensome to the university faculty researcher, but is nonetheless employed. For certain projects sponsored in this manner, certain intellectual property (IP) agreements often need to be agreed to and supported with written documentation with the industry or business concern that negotiates the main contract. The IP agreements are negotiated between the industrial or business organization and the university administration, generally the Office of Sponsored Research, or the university office that negotiates sponsored research agreements. The faculty member is generally not directly involved in the negotiation.

The US government is normally not involved with subcontracts to academic researchers when they are employed on contracts with industrial firms working as prime contractors. However, the industrial or business organization will generally require that the university agree to the same terms and conditions that are in the master contract. Also, for many projects sponsored by the DOD and some other agencies, the topic of the research may be export controlled under ITAR (International Trade in Arms Regulations) and/or EAR (Export Administration Regulations) restrictions. These restrictions are discussed in more detail in Chapter 8.

5.1.4 Grants

A grant is the most common (and most desirable) instrument used by the US government to fund research performed in academic institutions, particularly basic research provided to single investigators or small groups of researchers. A grant is basically a "gift," where funds are provided to perform a specified research project in a given area or topic. The intent of the grant is to advance the science and technology of the given subject. The research that is being supported is intended to be fundamental research (e.g., 6.1 research and/or 6.2 applied research) and is generally considered to be high risk, with potentially high payoff if the research proves successful. In program development the government program planners often consider the high-risk nature of the research, and make allowances for a certain fraction of the funded projects to be unsuccessful. The basic planning principle is that exploration into unknown areas and topics will not always produce immediate and tangible results. In fact, if a certain number of supported projects do not, in effect, fail, the program plan has not been well developed and sufficiently aggressive in exploring new areas. When a research project "fails," much is learned, and often a new and novel approach and path to a scientific and technical problem results and evolves, which can lead to new research funding opportunities. From this perspective, one can easily conclude that basic research never actually "fails" since much new knowledge is generated and learned.

Basic research is intended to address fundamental questions and to explore new directions in science and technology, and work in this area does not necessarily produce immediate hard output or results. The work is intended to help define new directions for future projects based upon fundamental scientific and engineering laws and principles. Projects supported by agencies such as the NSF and the NIH, as well as certain programs from other agencies, make extensive use of grants as a funding instrument, with the NSF and NIH making almost exclusive use of grants as a funding instrument for their research opportunities, with the exception of their large centers programs. Although the grant is essentially a gift, the funding agencies require that the recipients of the grant funds provide periodic progress reports and a final report. Oral presentations of the research program results at program review meetings, where government program managers and government research laboratory personnel can question methods and results, are expected and often required. The final reports required at the termination of a research grant are available to the general public and can be obtained from government database archives. It is also common for grant funding agencies to require dissemination of the research results into the public domain by mechanisms such as publication of the research results in the open literature, presentations at conferences and workshops, etc.

5.1.5 Cooperative Agreements

The US government sometimes wishes to have their scientific personnel participate in research that is performed at an institution outside of government, and to which the government provides support funds. Also, certain projects and research activities, such as large academic research centers supported with US government funds, require extensive interaction between US government personnel and the academic institution staff and researchers. For these cases, support funds can be provided by means similar to a grant through a funding instrument called a Cooperative Agreement (CA). The Cooperative Agreement differs from a grant in that it permits significant and substantial interaction between government program managers, scientists, and researchers and the recipient organization or institution. Cooperative agreements are used in a wide variety of ways, and are commonly used to establish certain centers and laboratories that are dedicated to providing research and support services to the agency providing the support funds. Cooperative agreements are rarely used for single investigator projects, although it is certainly possible for an individual researcher to participate in research supported under a cooperative agreement, particularly if your institution has a center or institute supported under such an arrangement. Cooperative agreements will specify in detail the extent to which government personnel will be required or available to advise, review, approve, or otherwise be involved with project activities. The cooperative agreement will also clearly state the conditions for any changes or requirements for more clearly defined deliverables.

5.1.6 Other Transactions

The "Other Transactions" (OT) instrument is essentially an open category, consisting of agreements that don't fit under the contract, grant, or cooperative agreements instrument definitions. This category can consist of a wide diversity of agreements between the government and an external party. It's most often used for specialized agreements with industrial or business providers. The OT instrument can vary, but is a handcrafted, specialized agreement, for the purpose of funding research and development, including developing prototype systems, etc. There is no specific definition for an OT agreement. The advantage in an OT agreement is that the government's formal procurement regulations and statutes do not apply and, therefore, the participating agencies have enhanced flexibility to enter into agreements with certain companies and commercial sources that have difficulty in complying with the government's procurement regulations. These OT agreements can be quite useful to the government when certain specialized and novel products are desired. The Other Transactions authority was initially enacted in the National Aeronautics and Space Administration (NASA)

Act of 1958 to support research and development performed in industry. Later, the OT authority was extended to the Department of Defense (DOD), Department of Transportation (DOT), Federal Aviation Administration (FAA), Department of Health and Human Services (HHS), Department of Energy (DOE), Transportation Security Administration (TSA), and Department of Homeland Security (DHS). The Director of the Office of Management and Budget (OMB) may authorize other federal agencies to use OT authority under certain circumstances. The OT agreements are not commonly used for government-supported research in universities or academic institutions.

5.2 Where Do I Find the Appropriate Funding Agency for My Research?

In the previous section we discussed the BAA as a primary method the US government agencies use to advertise their research interests to the public. We indicated that the BAAs are updated periodically as research interests change, and the BAAs can be issued to indicate general interests, in which case the BAA is issued for an extended period of time, or they can be used to indicate a short-term interest in a particular area or topic. Whereas the former may have limited funding available since the BAA indicates interest areas, not specific funding opportunities, the latter are generally associated with an immediate interest and need and, as such, a specified amount of funding associated with that particular BAA is generally indicated in the BAA.

5.2.1 Agency Websites

Although you're going to want to read and analyze the published BAAs, the best places to start your review of potential research funding opportunities are the various agency and funding office websites. Essentially all funding sources have easily accessible websites that list available funding opportunities, as well as much additional information concerning their programs. Information you'll find of interest includes the office structure, management, divisions, or directorates by scientific or engineering discipline, topical departments within the divisions or directorates including areas of responsibility, and, most important, contact information for appropriate personnel associated with the various technical areas. This information is your first exposure to making the appropriate contact to find a funding agent or representative that could potentially be interested in your research. Information concerning timing for proposal submissions, proposal requirements and evaluation criteria, and funding levels are also often included on the websites. The websites deserve your close attention to gain familiarity with the particular agency or office you wish to investigate as a potential source to fund your research.

5.2.2 Grants.gov

While the various agency and office websites are invaluable sources of information concerning grant funding opportunities, you'll also need to become familiar with the Grants.gov website. Most government agencies now require that proposals be submitted through this website. The Grants.gov website was developed over a decade ago and has become a major tool for researchers seeking US government support for their research.

A centralized system for listing research grant opportunities offered by the federal government was established in 2002 with the creation of the Grants.gov program and website. This program was included as part of the President's Management Agenda (Public Law 106–107), with the goal of assisting prospective researchers in searching for and identifying appropriate research grant opportunities. The Grants.gov program establishes a centralized system and location for researchers to easily identify federal government research funding opportunities. The office is managed by the Department of Health and Human Services (HHS), and operates under the auspices of the White House Office of Management and Budget (OMB). The Grants.gov program offers a very convenient and easily accessible electronic means of searching for grant opportunities offered by the participating government agencies through the website (http://www.Grants.gov) that includes research funding opportunities, program descriptions, and proposal submission information for over 1000 different grant programs offered through 26 federal research funding and granting agencies, thereby facilitating interaction and communication with the federal government. Along with the centralized location (website) the program simplifies the process for proposal submission by standardization of the information concerning the grant opportunities, employing a standard application format, and making use of a common submission procedure. Proposal submission is performed electronically using downloaded forms that can be filled in, auto-populated, and even error checked. You will need to register on the Grants.gov system but, once you have, you can then submit your proposals to any of the participating government agencies by using just one secure login, which is a significant simplification compared with the system prior to the Grants.gov program, when each agency had different proposal formats and varying proposal requirements. Also, the proposals are submitted electronically, which saves considerable time, reproduction costs associated with printing and copying a proposal, postage, and the time delay associated with mail delivery of hard-print proposals. Another attractive feature of the Grants. gov program is that you can perform custom searches for future opportunities from selected agencies you choose and have them sent directly to you electronically through email or RSS notifications.

The Grants.gov website will be one of your main resources in your search for research opportunities as you build your research program. You will need to become very familiar with the website, and this website will become one of your main resources for information concerning grant funding opportunities and a primary means for proposal submission.

By carefully reviewing the agency websites and Grants.gov you should be able to fairly and quickly narrow the potential US government opportunities for funding your research to a relatively few agencies where you will want to focus your search and efforts. The agency websites, in particular, need to be investigated in detail to identify the particular agency and division that supports research consistent with your interests. The agency websites give a much broader view of the research that they pursue and in which they have interests, although there may be no or limited immediate opportunities, and the Grants.gov website will give you details concerning immediate opportunities, which may also be listed on the agency websites. These two information resources are the most important tools you'll have available, certainly when you first start the process. However, this information is not complete and you'll need to learn how to make contact with appropriate program managers and funding agents. We'll address this issue in the next chapter.

5.3 Other Research Funding Sources

So far, we've discussed only US government sources for research funding. The reason for this is that the US government is the dominant source of research funding provided to academic institutions and, therefore, is the most important source with which you need to build a relationship. However, other institutions, such as state government research offices, not-for-profit and charitable foundations, and industrial and business organizations, also exist and offer research funding for certain areas and topics and should be investigated. If you perform research in the sciences or engineering, industrial sponsors, in particular, are potentially very important to your long-term research plan. For example, certain funding agencies, such as the NSF, actively encourage collaborations between academic researchers and industrial partners to be established. Often, NSF, DOD, and DOE research opportunities welcome support letters from appropriate industrial sources. These sources don't necessarily need to provide any support funds, but letters from industrial collaborators indicating their interest in the research that is being proposed can be included in the proposal, usually in an appendix or attached to the proposal. These collaboration support letters may indicate material support, or access to industrial laboratories and experimental equipment not available at your university. The experts that populate review panels read the support letters, and the support letters are considered in the evaluation of the proposal as an indication of the importance and appropriateness of the research.

Also, certain funding opportunities, such as the Small Business Innovative Research (SBIR) and the Small Business Technology Transfer (STTR) programs often require the collaboration of academic researchers and industrial scientists and engineers. The STTR program, in particular, requires the collaboration between private business organizations and research institutions. These relationships and collaborations can be excellent sources of research funding to build your program. Therefore, you are going to want to identify industrial collaborators. Industrial sources, of course, are also a potential source of research funding. The level of available funding is generally lower than available from government sources, but the funding can be very effective in building your research program since the funding will usually be very "tuned" to exactly what research you are pursuing, and in which the industrial sponsors have great interest. In addition, other support, called "in-kind" is generally available and easy for the industrial or business organization to provide. This support is generally diverse, and can include a wide range of material and support items and services, including items such as access to laboratories and related equipment, access to device or component data not readily available elsewhere, exposure to information concerning practical and theoretical problems experienced in the real-world practice of engineering new devices and systems, access to fabrication and measurement support, and support of internships for students, etc.

Exposure to problems encountered by the industrial scientists and engineers is particularly important since these problems are probably what initiated the collaboration in the first place. That is, industrial collaboration interest in your work is probably motivated by your work offering the industrial or business scientists and engineers an explanation or solution to a problem they've encountered. This gives you a great opportunity to build the collaboration, with the very positive result of beneficial outputs for both of you. This type of support is incredibly important over the long run, and can quickly lead to state-of-the-art results, and publications. The industrial collaborations that are established have the advantage of not being limited by the typical government grant funding time limits, and can exist over many years, sometimes extending over decades. In my own experience, I've had collaborations with certain industrial scientists and engineers that have existed throughout my employment at three different universities, while several of my industrial collaborators have changed job locations two, three, and more times. The locations change, but the collaborations continue. Personal relationships are exactly that, personal relationships. You need to build as many of these relationships as practical. We'll address how to identify and build relationships with potential industrial collaborators in the next chapter.

A variety of foundations that provide research funding also exist. A foundation is a non-profit organization formed as a corporation or charitable trust and funded by private sources, such as a wealthy individual, family, or corporation, using money they donate. Examples include the Bill and Melinda Gates Foundation, the Ford Foundation, The William and Flora Hewlett Foundation, The Rockefeller Foundation, and many, many more. Altogether, there are on the order of 100 000 grant-making foundations established. Many of these have been created to support activities, including research and development, on very specific and focused topics. In order to see if any of these foundations make grants in your specific specialty you'll need to do some research. The website for the specific foundation in which you may have an interest is always the best place to start. To help you in identifying a specific foundation to contact, there are two powerful electronic resources available offered by the Foundations Center (www .foundationscenter.org), which maintains a database of grant-making foundations both in the USA and on a global basis. The *FC Search* and the *Foundation Directory Online* services offered by the Foundations Center include a Fields of Interest Index that will help you sort the available foundations that are potential grant sources for your research. Once you identify a foundation as a potential funding source for your research, you should contact them informally before submitting a proposal. The information contact can consist of a phone call or letter of inquiry to determine their interests in your research, and the availability of funding. If there is a potential interest they may request a short, one or two page, "white paper" in which you briefly outline your research, what you want to pursue, and how the work may be of interest to the foundation. Informal contact is always desirable before going to the time and effort to write a full proposal, even if there is a published call or request for proposals

5.4 Research Funding Background and Trends

The question of which funding agency or source offers the greatest potential for funding your research always arises. In order to address this question, it helps to understand how research budgets are allocated among the various offices and organizations. A famous quote is attributed to bank robber Willie Sutton, who, in response to a reporter's question "Why do you rob banks?", is rumored to have responded "Because that's where the money is!" Sutton later claimed he never made the statement, which he attributed to the overly creative reporter, but the quote stands and contains a valuable lesson, particularly for researchers searching for support funds for their research. The lesson, slightly expanded and simply stated, is: Focus your search (at least your initial search) on the agencies and grant-funding offices that have budgets where you'll have the greatest potential for success. In order to help you understand the research budgets, it's helpful to review the funding trends over time for research support for several popular sources of research grant funding.

Funding trend data are collected for the US government by the NSF and published in their National Patterns of R&D Resources series and available at the NSF website http://www.nsf.gov/statistics/natlpatterns/. The American Association for the Advancement of Science (AAAS) (http://www.aaas.org) regularly analyzes funding trend data, using a variety of data sources, and publishes graphs showing various trends. The data presented in this section is sourced from the AAAS Historical R&D Data from their website (https://www.aaas.org/page/historical-trends-federal-rd), and the graphs in the figures are reprinted with permission.

Figure 5.1 shows the expenditures in billions of constant 2015 fiscal year (FY) dollars of the total national R&D performed by various

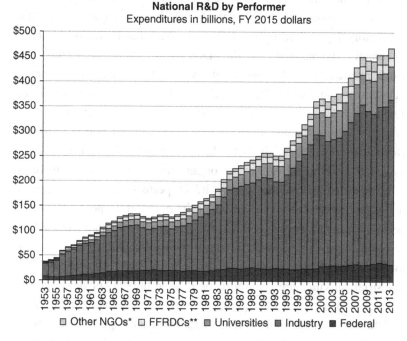

Fig. 5.1 National research and development (R&D) indicated by performer (AAAS). Symbols: *, Non-Government Organizations; **, Federally Funded Research and Development Centers. Source: National Science Foundation, *National Patterns of R&D Resources* series. Constant-dollar conversions based on GDP deflators from *Budget of the U.S. Government FY 2016.* © 2015 AAAS.

organizations and how the funds track over time, dating back to 1953. The dark gray markers indicate the funds provided for R&D performed by the federal government at US government owned and operated laboratories, such as the laboratories operated by the DOD, DOE, NASA, etc. The other markers indicate the funds provided to industrial laboratories, university laboratories, Federally Funded Research and Development Centers (FFRDCs), and other non-government organizations (NGOs), respectively. The FFRDCs are independent R&D centers that are managed and operated by non-government contractor organizations to perform research and development activities for the US government. They are fully supported by the US government and are not permitted to compete for announced research solicitations.

There are currently 39 FFRDCs sponsored by the US government. They include organizations, for example, such as Argonne National Laboratory (sponsored by the DOE), the National Renewable Energy Laboratory (sponsored by the DOE), the National Center for Atmospheric Research (sponsored by the NSF), the MIT Lincoln Laboratory (sponsored by the DOD), the California Institute of Technology Jet Propulsion Laboratory (sponsored by NASA), and the Frederick National Laboratory for Cancer Research (sponsored by the Department of HHS and the NIH). Each of the 39 FFRDCs is sponsored by a US government agency or organization. As indicated, there is a change in the percentage of R&D performed by the various organizations over time. The vast majority of research and development performed in the nation is done by industry, with about 70% of all research currently performed by industrial organizations. However, this includes everything classified as research and development, and the industrial activities are heavily focused upon the "D" or development activities. Industry, in fact, performs only a limited amount of basic scientific or engineering research, which we'll see is performed mainly in university laboratories. It is interesting to note that the percentage of R&D performed by US government laboratories and by the FFRDCs has been decreasing over time, while the percentage of R&D performed by industry and universities has been increasing.

The issue of who provides the support funds for national R&D activities is addressed in Fig. 5.2. The plot indicates the level of funding provided, indicated in billions of constant 2015 fiscal year dollars. This measure permits the effects of inflation and the changing value of the dollar over time to be normalized from the calculation. In this manner the graph indicates the "true" distribution of the funds and permits an accurate assessment of changes over time. As the plot indicates, the trend has been for the amount of R&D funds to increase over time for the three main sources, the federal government, industry, and other organizations, which includes sources such as foundations, gifts and donations, etc. While all three sources have increased their funding provided for national R&D, the greatest increase is for industrial-funded R&D, which has increased at greater than twice the rate of

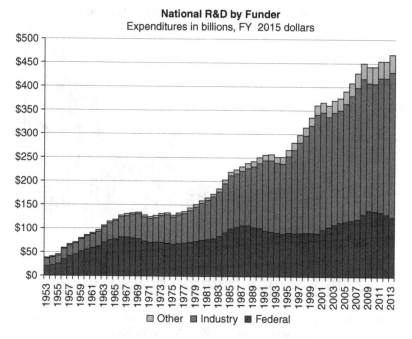

Fig. 5.2 National research and development (R&D) indicated by funding source (AAAS). Source: National Science Foundation, *National Patterns of R&D Resources* series. Constant-dollar conversions based on GDP deflators from *Budget of the U.S. Government FY 2016.* © 2015 AAAS.

federally funded R&D. However, the majority of the industrial-supported R&D goes to support R&D activities in their own organizations, with heavy emphasis upon development, and limited support for basic research. The amount of funding provided by industrial sources to universities for academic research has historically been limited, and remains so to the current time. However, industrial support is still an area that should be pursued and can be a very strong component of a successful academic research program.

The divide between defense-related and nondefense-related R&D funding provided by the US government is shown in Fig. 5.3. As previously indicated, prior to 1960 essentially all R&D funding provided by the US government was for defense-related work. This stemmed from the positive results of government funded R&D and the

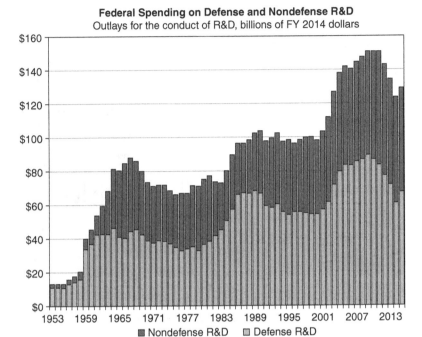

Fig. 5.3 Distribution of national spending for research and development (R&D) listed by defense and nondefense sources. Source: AAAS, based on OMB Historical Tables in *Budget of the United States Government FY 2015*. FY 2015 is the President's request. Some Energy programs shifted to General Science beginning in FY 1998. © 2014 AAAS.

partnerships that were developed between the US government and industrial and academic research efforts during and following the Second World War. With the onset of the Cold War there was a large increase in federally supported R&D activities, both for defense-related and nondefense-related R&D work. The nondefense-related R&D was primarily for health-related R&D, although other nondefense activities also saw increased federal funding. While the amount of federally provided funding has varied over the years, there has been a gradual shift to a larger percentage of the federal budget being provided for nondefense-related R&D, and at the present time, there is near parity with about half the authorized R&D funds going to each category.

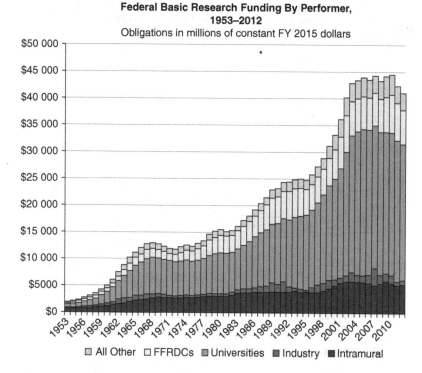

Federal Basic Research Funding By Performer, 1953–2012
Obligations in millions of constant FY 2015 dollars

☐ All Other ☐ FFRDCs ◼ Universities ◼ Industry ◼ Intramural

Fig. 5.4 Federal basic research funding listed by performer (AAAS). Source: NSF, National Center for Science and Engineering Statistics, *National Patterns of R&D* series, based on national survey data. The FY 2012 data are preliminary. © 2015 AAAS.

The data shown in Figs. 5.1–5.3 refer to the total national R&D activities in all categories. While Fig. 5.2 indicates that industry performs about two-thirds of all R&D within the nation, the situation is significantly different when basic research is considered. Basic research is, of course, the main category of interest to academic researchers. The data shown in Fig. 5.4 indicate the amount of basic research funding provided by the US government to various R&D performing organizations. The government provides about 15% of the federal basic research funds to government-owned laboratories and a very small percentage to industrial laboratories, while the majority of the basic research funds, about 60%, are provided to academic researchers, with the remaining

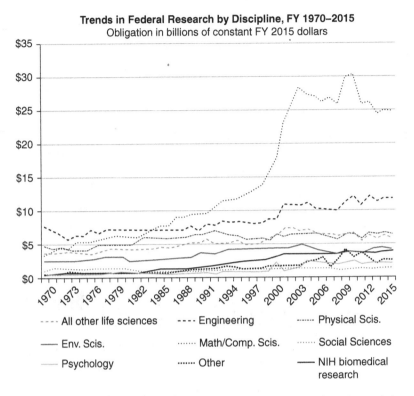

Trends in Federal Research by Discipline, FY 1970–2015
Obligation in billions of constant FY 2015 dollars

---- All other life sciences ---- Engineering ------ Physical Scis.

—— Env. Scis. ······ Math/Comp. Scis. ······ Social Sciences

—— Psychology ······ Other —— NIH biomedical research

Fig. 5.5 Distribution of federal research and development (R&D) funds by receiving discipline (AAAS). In this figure, "Other" includes research not classified (includes basic research and applied research; excludes development and R&D facilities). Life sciences are split into NIH support for biomedical research and all other agencies' support for life sciences. Source: National Science Foundation, *Federal Funds for Research and Development* series. FY 2014 and 2015 data are preliminary. Constant-dollar conversions based on OMB's GDP deflators. © 2015 AAAS.

25% going to FFRDCs and other performers. As this graph indicates, the primary recipients of federally sponsored basic research funds are academic researchers and laboratories. This indicates, of course, that your primary source of research funds to build your research program is the federal government, through the various funding agencies and offices.

Another interesting question relates to which disciplines receive the federally sponsored R&D funds. This question is answered in the data shown in Fig. 5.5. This graph shows the trends in federal research distribution over time by various disciplines. Dating back to 1970 there has been a slight increase in federal research funding for the various disciplines, with a slightly increased growth rate for funding for engineering, and a significant increase in funding for biomedical research sponsored through the National Institutes of Health (NIH). The doubling of the NIH research funding budget from about $13 billion to $26 billion over the five-year period of 1998 to 2003 was a result of an act of Congress. Since this time the NIH research budget has slightly decreased over the past decade, with sizeable variations from year to year.

The distribution of the federal support by government agency for university research over time is shown in Fig. 5.6. The data are shown through fiscal year 2013. The big increase in funding for 2009 is a result of the American Recovery and Reinvestment Act (ARRA) of 2009, commonly referred to as the Stimulus or The Recovery Act, which provided government funds to create jobs and stimulate the economy following the 2008 economic collapse. The Act provided $7.6 billion for scientific research, divided between the National Science Foundation (NSF), the Department of Energy (DOE), NASA, National Oceanic and Atmospheric Administration (NOAA), the United State Geological Survey, the National Institute of Standards and Technology (NIST), and the National Institutes of Health (NIH). The US Department of Defense (DOD) did not receive any ARRA scientific research funds. The increase in research funding was a one-time event and, for the years following 2009, research funding has, in fact, declined. However, the relative proportion of federal support provided for university research and development across the various agencies has remained essentially constant over time.

As indicated in Fig. 5.6, by far the largest amount of research funding to universities is provided by the Department of Health and Human Services (HHS) through the NIH. The overall increase in university research funding provided by the federal government over the years from 1999 to 2003 was due to the NIH doubling act passed by Congress. Funding from the other agencies was essentially constant over this period. For the period after 2003, research funding, in fact,

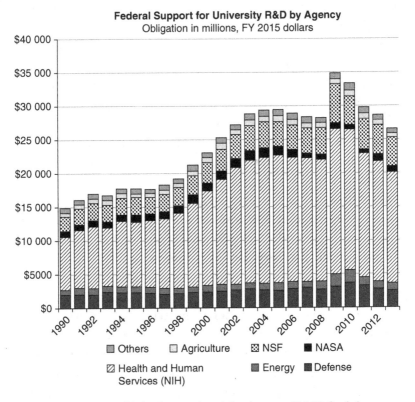

Fig. 5.6 Distribution of federal research and development (R&D) funds by agency (AAAS). Source: National Center for Science and Engineering Statistics, *Federal Science and Engineering Support to Universities, Colleges, and Nonprofit Institutions* series, based on national survey data. Includes R&D and R&D plant. FY 2009 and FY 2010 include Recovery Act funding. © 2015 AAAS.

declined, primarily due to a decrease in funding through NIH, but also due to a reduction in funding provided by NASA. Research funding provided by the US Department of Defense (DOD) has demonstrated a gradual increase over the period from 1990, but represents a minor fraction of the federal research support funds provided to university research. However, as a mission agency, the DOD supports only research in select areas consistent with the DOD mission and, in certain disciplines, such as electrical engineering, the DOD is a major source, accounting for around 70% of all academic research funding in this field.

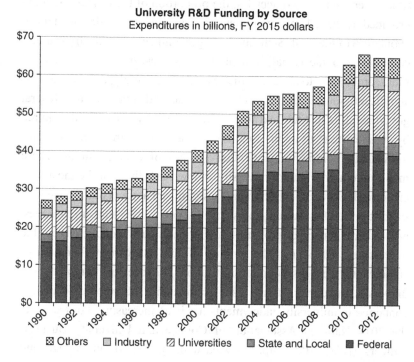

Fig. 5.7 Support for university research and development (R&D) indicated by source of funding (AAAS). Source: NSF, National Center for Science and Engineering Statistics, *Higher Education R&D* series, based on national survey data. Includes Recovery Act funding. © 2015 AAAS.

The NSF remains a significant, and growing, source of academic research funds. It is clearly second only to the NIH in the amount of funds provided to academic research. Also, since the NSF is not a mission agency, it can support virtually any basic research topic area, although the research topic needs to be identified with an NSF directorate and division.

Universities receive research funding from a variety of sources, as indicated in Fig. 5.7. The federal government is the single most important source of research funding, providing about two thirds of the total research and development funds expended on research activities in US academic institutions. Although there has been a significant increase in research expenditures for university research, increasing about 2.5

times over the past two decades, the proportion of the research funding provided by the federal government has remained essentially constant, at about 60% of the total. State and local government and industry sources, combined, provide a small amount of the total research and development funds, amounting to about 5% of the total. Surprisingly, the second most significant source of research funds, second only to the federal government, is institutional funds. These funds are raised and acquired by universities through fund-raising activities from foundations, individuals, companies, and other sources of donated funds. These monies are used to build university endowments that, for the most part, are locked in investments, although a significant amount of the generated interest is used for chaired professorships, new faculty start-up packages, and other sponsored program-building activities conducted by the universities. Of course, the amount of the endowment funds varies significantly from institution to institution. However, as more and more universities undertake fund-raising campaigns, the amount of money contained in university endowments has been increasing. For example, the 2014 National Association of College and University Business Officers (NACUBO: http://www.nacubo.org) report on college and university endowments contains a survey of 847 colleges and universities, and concludes that, collectively, university endowments total about $455 billion. As of 2013 there were 83 universities that reported an endowment exceeding $1 billion. The fund-raising activity is responsible for the increase in the institutional funds indicated in Fig. 5.7.

Laboratory facilities, research equipment, and instrumentation are fundamentally important to academic research. Experimental research requires significant investment in facilities, which can be very expensive both to obtain and then to maintain. Also, experimental research, if it is to be performed at the state-of-the-art, requires continual upgrades of equipment and instrumentation. Also, theoretical research generally requires computational facilities, computers, and computing-related equipment. Appropriate laboratories and related equipment and instrumentation are difficult for academic researchers to acquire due to high cost and limited sources of funds. In order to address this need, some government funding agencies, such as the NSF (through its Major Research Instrumentation (MRI) program), and the US DOD (through its Defense University Research Instrumentation Program (DURIP)), sponsor grant competitions for major research equipment, generally on an annual basis. These competitions are directed towards major equipment items that are difficult to acquire through the normal grant process, and usually involve equipment items with high price requirements, which can be up to or over a million dollars.

The US government also provides funding for equipment and related items. Although the amount of funding for laboratory facilities, equipment, etc., varies over time, the federal government provides about 20% of its total investment to their own laboratories (distributed among the various agencies), about 50% of its total investment is provided to industrial laboratories, and slightly over 15% is provided to academic laboratories, and the remaining funds are provided to FFRDCs and other performers. Excluding the ARRA funding years, the funding trend for equipment has been essentially constant for the past decade.

The normal grant process generally does permit the acquisition of smaller equipment items that are specifically required to perform the proposed research described in the grant application. However, only limited amounts of funds are normally permitted to be included.

5.5 What Time of the Year Should I Submit My Proposal?

The correct time to submit your proposal to a potential funding agency is a very important issue. The old adage is: "The time for cookies is when cookies are passed!" This means the best time to submit a proposal is when the sponsoring agency or organization has funding available to support your research. This statement is very, very true for all funding sources, and particularly when responding to specific calls for proposals (RFPs, CFPs, etc.). When a specific topic RFP is published, it will generally indicate a level of funding expected to be provided for the overall program, as well as limits for individual proposals. The RFP resulted from strategic planning performed by the funding agency program managers or program directors and a portion of the agency research budget was allocated to the program. This means that the money is real, and will be available and provided for research and development in accordance to successful proposal requests. The published RFPs represent real opportunities that need to be pursued. A downside is that the published RFPs often occur at very inconvenient times for faculty members, particularly those heavily engaged in other activities. Time management, in these cases, becomes critical. In order to respond to the funding opportunity, it may become necessary to drop other activities in order to prepare a proposal for submission. To be successful, you'll need to learn to manage your time commitments, which means you'll most likely need to learn to turn down certain invitations to get involved in activities that you'd like to pursue, but which conflict with proposal preparation and research activities. By all means, do this! Remember your prime mission is to build a research program, and this requires funding support, which can be acquired only through proposal preparation and submission.

As indicated earlier in this book, there are two main types of funding opportunities, the specific calls for proposals (RFPs), and the open submission windows. There is considerable variation in these opportunities offered by different agencies and funding sources. The NSF, for example, generally has two annual "window" opportunities to submit "unsolicited" proposals. There is one window in the fall, generally in the October/November timeframe, and one in the spring (generally in February). However, some directorates and divisions accept proposals only in the fall window and do not have the spring window. The specific window requirements and dates vary by directorate and division, so you'll need to check their website to determine the particular due date for your proposal.

The US Department of Defense (DOD) will accept unsolicited proposals for research essentially any time of the year. Their research interests are indicated in the BAA published by each specific agency and, under the BAA, they can accept proposals from any qualified organization. While this may sound more open than the defined submission windows offered by NSF, be aware that the agency to which you wish to submit your proposal may not have any available budget to support your research, even if they have interest in your work. For this reason, you need to contact the appropriate program manager before actually submitting your proposal, both to determine their interests in your research, and to determine if funding is available. The best time to approach DOD funding agencies for research support is at the beginning of the fiscal year, when they have received their research budget for the year, and at the end of the fiscal year, when they may have research funds remaining that have not yet been committed.

The US government fiscal year runs from October 1 to September 30. Funds that remain at the end of the fiscal year are generally called "year-end funds" and, when available, offer a good opportunity for the program managers to initiate new research projects in which they have interest. The remaining funds result for several reasons, but generally they come from a gap between committed and available funds. Program managers and program directors want to spend these funds because, if not spent, they will be returned to the agency, and the program manager's research program funds could potentially be reduced by the amount returned. To avoid this, many program managers will offer the remaining funds to new researchers, or to those with novel ideas, that were not funded in the normal process. Generally, these year-end funds are provided only on a limited basis for a one-year performance period, but offer an opportunity to demonstrate your research to the program manager, which can lead to much more significant funding for additional research. We'll address this issue later when we discuss how to work with funding agents. Other agencies and funding sources function essentially in the same manner as NSF and DOD and you'll need to learn the specific opportunities from their published information. The best place to start, again, is their websites, which will include information on proposal submission requirements, etc.

You need to be aware that published requests for proposals for research on specific topics always have a clearly defined submission date with a clearly defined submission time. This date must be respected... no exceptions! As a US government program manager, I've witnessed numerous proposals declined simply because they weren't submitted by the specified time on the specified due date. This may seem unduly harsh to a person who has just spent many hours and days working on a proposal. However, most agencies take a hard stand on this issue, and if you miss the submission time and date, your proposal will, most likely, be returned without review. Of course there are always exceptions to the general rule, and there are some circumstances under which a late proposal may be accepted, but these conditions are rare, and it's not worth the gamble of needing to appeal your case. It's much easier to simply meet the proposal submission deadlines.

5.6 What We've Learned

In this chapter we've discussed five fundamental principles associated with US government research funding. Included in the discussion are explanations of the different legal instruments that the government has available to transfer funds from the various funding agencies to research performing organizations, as well as explanations of the procedures and techniques that funding agencies use to advertise their research grant funding opportunities. In particular, the need to make use of the funding agency websites, and the Grants.gov website were specifically addressed. We also reviewed research funding trends over time, and noted which organizations and funding agencies provided the greatest amount of research funding. We concluded the chapter with a discussion of the optimum time of the year to submit a research proposal.

6 Making Contact and Communicating with Program Managers and Program Directors

An important question most young faculty members ask relates to how the proposal management process within government agencies is conducted, and who is responsible for the execution of the process. The mission agencies term the people who manage the research proposal process Program Managers, while the National Science Foundation (NSF) terms them Program Directors. Sometimes the general title of Program Officer is used. The duties of the program managers and program directors are similar, but fundamentally different in authority, and the manner by which the proposal management process is performed. Perhaps one of the first questions you might ask is: "Do I need to make personal contact with funding agency program managers and program directors?" The answer varies somewhat, depending upon the funding agency you wish to solicit for funding, but the general answer is an unqualified "Yes, you are well advised to make personal contact with the program managers!"

6.1 The Need to Personally Communicate with Program Managers and Program Directors

For the mission agencies, such as the offices within the Department of Defense (DOD), the Department of Energy (DOE), NASA, etc., personal contact can be paramount. Program managers in these agencies and offices generally have discretion and authority to provide funding to researchers for proposals they deem worthwhile, sometimes without the need to conduct in depth reviews or evaluations of the proposals. Although the mission agencies make use of external or internal proposal review panels staffed with technical experts, the final decision as to which proposals to select for funding rests with the program manager, although they generally accept the review panel results. However, in their core program, the program manager selects the proposals to be funded, and any reviews solicited by the program manager are used as guidance, and the results of the review do not necessarily determine the program manager's decision as to whether to accept or reject the proposal for funding. The program manager will often personally read and evaluate proposals and form his or her own opinion of the quality and importance of the proposed research. They have the authority to disregard reviews that they feel don't accurately represent or value the proposed research. The program manager's evaluation of the proposed research, and exactly how it supports his or her research program objectives, will dominate the decision.

In order for the program manager to select your proposal for funding, he or she needs to have confidence both in you as a potential principal investigator and in your ability to perform quality research, as well as in the specific topic being proposed. The program manager will evaluate the prospects for success, and how the proposed research will support the overall goals they are pursuing. Therefore, they need to meet you and give you a chance to present your research pursuits. An unsolicited proposal submitted blindly, without prior contact with the program manager, has little chance of being selected for funding.

However, for non-mission agencies, like the NSF, personal contact is useful, but not absolutely necessary since your proposal will be evaluated by a panel of experts who come from institutions outside of the NSF. Program directors or other experts within the NSF do not personally evaluate proposals and do not generally enter into the evaluation and discussion of a proposal during review panel deliberations. The results of the panel review will determine which proposals are selected for funding. Therefore, the review panel ratings of all proposals are extremely important, and the NSF program directors can only recommend a proposal for funding if it has received high ratings. The NSF program directors do not actually have the authority to obligate the NSF to fund a particular proposal, and their role is to organize and manage the review process, usually by means of a proposal review panel, although they will sometimes send certain proposals to experts outside of the NSF for review. They then rank/order the proposals according to the review panel or outside reviewer evaluations, and then they make recommendations to the division director regarding the proposals to be funded. The actual selection is the responsibility of each division director, although they normally approve proposals for funding by concurring with the program director's recommendation. If they do not concur with a proposal selection recommendation, that particular proposal will not be selected for funding. This occurs only for well-defined reasons that are entered into the official records.

Personal contact with mission agency program managers has always been important, and is becoming more so as research budgets become stressed. Simply stated, you need to demonstrate to the program manager that you are performing research that is contributing advances to an area in which they wish to invest. The program manager needs to have confidence both in you as a researcher, and in the technical area in which you work. I emphasize that the program manager working in a mission agency, as stated above, personally makes the decision of which researcher will be supported and provided a research grant. For a non-mission agency such as the NSF, the program director will make their recommendation decision from a ranked list of proposals evaluated by an expert panel. In either case, the program manager or program director is an important person in making, or recommending, the decision regarding grant acceptance and funding, so it's a very good idea to meet them and establish a personal relationship. If you survey faculty members who have been successful in obtaining research funding over a sustained period of time, the one common characteristic you'll discover is that they all have established personal relationships with program managers and program directors. Generally, these personal relationships form networks that endure over many years, and sometimes decades. The relationships are beneficial to both sides, and establishing a personal relationship with a program manager that may grow and extend to other program managers will be one of your main priorities. In order to understand why this is so, it is helpful to understand how program managers interact within their home agency and colleagues, and how the program managers are evaluated in their performance within their agency.

6.2 Program Managers and Program Directors

United States government program managers and program directors come from the research community. At the majority of the US government funding agencies, the people who are selected to serve as program managers will be PhD-level scientists and engineers. In some instances, the program managers may not hold a PhD, but they will have significant professional experience, which gives them expertise and practical knowledge essentially equivalent to their PhD colleagues. All program managers and program directors are highly trained and are experts in their fields. Their responsibility within their agency is to learn and understand research trends and developments within science and engineering, and particularly within their specialty field. They are expected to keep their knowledge up to date and to know and understand the state-of-the-art in their areas. In order to accomplish this, they have numerous approaches available to them. For example, program managers will study the technical literature, including professional publications and technical reports from a variety of sources, they will attend technical and professional conferences and meetings, they routinely visit research laboratories and discuss research trends and developments with leading researchers in academic, industrial, and government organizations, and they read and evaluate many research proposals, serve on research proposal evaluation panels where various approaches to research are discussed, and host meetings with researchers who will travel to meet them for the purpose of presenting their research approach and results.

Program managers will periodically travel to visit scientists and engineers, particularly those they are supporting, to review their progress and recent developments. The program managers also interact extensively with research scientists and engineers both within their own organization, and within other government organizations. It is not unusual, for example, to see research scientists and program directors from organizations such as the US Army Research Office, the Army Research Laboratory, the US Office of Naval Research, the Navy Research Laboratory, the Air Force Office of Scientific Research, the Air Force

Research Laboratory, DARPA, the Department of Energy, NASA, the NSF, and others, all attend, with outside scientists and engineers, meetings on research and development topics of mutual interest. Technical meetings are often organized and hosted by government program managers, with the financial support of their organizations, for the purpose of examining recent developments in a given area, and to explore approaches for future research directions. Strategic planning meetings occur on a routine basis, and many of these planning sessions result in research funding opportunities for the academic and industrial communities.

In this process, the program managers and program directors become aware of what scientific and engineering approaches are being followed, what results are being achieved, what future trends are developing and, most importantly, the identity of the major research performers that present the most novel and promising approaches and results. The majority of these meetings are open to attendance by academic and industrial scientists and engineers, and attendance at these meetings can help a researcher place their research approach in perspective, to be exposed to alternative approaches, and to help in future planning. Attendance at these meetings can also help make you aware of which program manager may be a potential source of support for your research, and your presence at the meeting gives you the opportunity to make contact with these program managers in an informal setting. While it will not be possible to attend all these meetings, you should make an attempt to attend those that are directly pertinent to your research topic. These meetings are generally widely and publicly announced and advertised. Again, agency websites are a good location to search for information.

6.2.1 The Program Manager Evaluation Process

Mission agency program managers and NSF program directors are evaluated in different ways, although both go through detailed annual performance reviews. In order to better understand the role of the program manager or program director, it's helpful to discuss how they are evaluated and rated by their organization. This, in turn, helps you to understand what they are looking for in a new researcher, and how you can best approach them and introduce and explain your research ideas and goals to them. Also, understanding how the mission agency program managers are evaluated places in perspective the reason that they have authority to select certain proposals for funding, and why the proposal merit review process does not necessarily restrict them in the same manner as it does for program directors at NSF.

Program managers and program directors working for US government funding agencies go through robust evaluation procedures. They, along with their colleagues and managers, have the responsibility for determining the research areas in which their agencies will invest research funds. Program managers also determine which researchers actually receive research funds in response to their submitted proposals, while NSF program directors serve a similar purpose, and make recommendations for which proposals are selected for funding. Both have the responsibility of selecting the successful proposals from competing proposals, which can be a large number. In this regard they have significant control over the direction of national research, and play a very critical role in the direction that science and technology emerges and develops. For this reason, the program managers and program directors go through a very robust and detailed review and evaluation procedure, generally on an annual basis.

During these reviews the mission agency program managers are required to stand before their department and agency managers, directors, and colleagues, and sometimes outside experts, and explain and defend their program, detailing what they are attempting to accomplish, progress that has been achieved, how the research supports the agency mission, and whom they are supporting with research grants. Program managers have a technical area that they oversee that is usually quite well defined and specific, with well-stated research goals. They generally have a good idea of what research areas and technical subjects they want to support and have a well-developed and formulated strategic plan. They are generally trying to accomplish an end result that has been defined by their in-depth knowledge of a given area, and through meetings and discussions with their colleagues and managers. Often, the end goal will consist of a larger overall view of a specific area and may include development of a new approach to a complex problem. They may be coordinating their research program with that of their colleagues from within their own agency, or those from other agencies. In order to accomplish their goal, they will need advances in a variety of scientific and technical subjects, all related to the end goal. In their program review the program managers will generally present and describe their view of what they're attempting to accomplish, along with a description of the work the researchers to whom they are providing research grant funding are pursuing, the progress that has been achieved, and the future directions for the research.

In preparation for these reviews the program manager will generally solicit input and results from the researchers they sponsor. In their overview of their programs, the program managers will also address how their research program aligns with their agency mission and goals, and what future trends are developing. They will present and discuss the research areas that are developing and why their agency should be investing research funds in these areas. Also, in the program reviews, the program managers will often present a list of specific accomplishments, publications, technical or scientific performance awards received, etc. The technical or scientific performance awards, along with any they personally may have received, include those received by the researchers they have sponsored. In particular, all program managers enjoy announcing that a researcher, and particularly a young and new researcher that they are supporting with a research grant, has received a recognition award from a professional society, etc. This is considered an indication that the program manager is sponsoring high-quality, significant, and important research, and the recognition award is a positive factor in the program manager's review and evaluation. To be mentioned in these reviews is also a very significant recognition for the researcher, and helps to make their research known to other program managers in attendance at the review. Program managers are always on the lookout for a new, young researcher that appears to be a developing research talent. Your goal should be to become this new talent waiting to be discovered!

Annual performance reviews for NSF program directors are not as detailed or complex a process. The performance evaluation is conducted between the program director and the division director, and the program director will complete a standard evaluation form that requires that certain specific topics be addressed. The form is submitted to the division director and serves as the basis for the annual performance review.

6.3 Funding Agencies and Organizations

Another important question most, particularly new, faculty members just starting a research career, ask is: "What funding agency should I approach, and do they all function in the same manner?" There are a variety of funding agencies for you to investigate. The best places to start are the funding agencies with the largest research budgets for external grants. That is, although some agencies have relatively large research budgets, some keep most of their research funds for internal use and support of their agency research laboratories. As the data presented in Fig. 5.6 indicate, the federal agencies that provide the largest amount of financial support for university research and development are, in rank order, the National Institutes of Health (NIH), the National Science Foundation (NSF), the Department of Defense (DOD), the Department of Energy (DOE), and the National Aeronautics and Space Administration (NASA). All of these agencies have established research offices to manage their external research. These five main funding agencies are briefly described below.

6.3.1 The National Institutes of Health (NIH)

The National Institutes of Health (NIH) is a mission agency that supports science in the areas of biology and the behavior of living systems, with the goal to apply that knowledge to extend human life and reduce illness and disability. The various NIH Institutes and Centers (ICs) provide funding for a wide variety of programs, and the NIH uses activity codes to differentiate the research-related programs. For example, "Series" codes for research-related activities are: Research Grants (R series); Career Development Awards (K series); Research and Training and Fellowships (T & F series); and Program Project/Center Grants (P series). A research grant to support a well-defined and specific research project is called an R01 grant. This is, in fact, the most common grant program offered by the NIH. There is no dollar limit to the grant, unless a specific limit is indicated in the Funding Opportunity Announcement (FOA), which is the NIH's version of a Call for Proposals. However, advance permission from the NIH is required for any budget request in excess of $500 000 in direct costs for any year. The NIH R01 grants are typically awarded for a performance period of three to five years. The NIH Institutes and Centers periodically publish FOAs, either as a Program Announcement (PA) or as a Request for Applications (RFA), and these are generally open for a period of time ranging from one to three years for proposal submissions. The NIH Institutes and Centers will also accept unsolicited proposals for research that is not appropriate for the research described in their published FOAs. Unsolicited proposals should be submitted through what NIH terms "parent announcements," which are funding opportunity announcements that cover the entire breadth of the NIH mission.

The NIH seeks to support high-quality research that is relevant to public health requirements and research that is consistent with NIH Institutes and Centers priorities. Proposals submitted to the NIH go through a dual peer review process, which is mandated by statute (Section 492 of the Public Health Service Act). Each proposal is first reviewed by a Scientific Review Group, which is mainly made up of non-federal scientists with expertise in the relevant scientific discipline of the proposals they are assigned to review. The second review is performed by NIH National Advisory Councils or Boards, which are made up of both scientific and public representatives chosen for their expertise, interests, or activities in appropriate areas related to the proposed research. Proposals must be recommended for approval by both levels of review before they can be recommended for funding. If you wish to pursue NIH funding opportunities, it is always best to contact the appropriate person within an NIH Institute or Center to discuss your research before submitting a proposal. Information concerning NIH interest areas and funding opportunities at the NIH can be found at the NIH website (http://www.grants.nih.gov), and funding opportunities are listed on the Grants.gov website.

6.3.2 The National Science Foundation (NSF)

The National Science Foundation (NSF) is an independent federal agency established by an act of Congress in 1950 "to promote the progress of science; to advance the national health, prosperity, and welfare; to secure the national defense...." It is a non-mission agency dedicated to supporting basic research in fundamental science, engineering, and education, and is the only federal agency that provides support for research in all fields and areas of fundamental science and engineering, except medical sciences. The NSF's goal is to support research that is high-risk, but potentially high pay-off, as well as to support novel collaborations and approaches. While providing about a quarter of the total external research funding to US colleges and universities, the NSF provides the majority of the research funding in areas such as mathematics, computer science, and the social sciences. For example, about 80% of the external funding for computer science academic research is provided by the NSF, primarily through the CISE directorate. The NSF is composed of seven technical directorates (the Directorate for Biological Sciences (BIO), the Directorate for Computer & Information, Science & Engineering (CISE), the Directorate for Education and Human Resources (EHR), the Directorate for Engineering (ENG), the Directorate for Geosciences (GEO), the Directorate for Mathematical and Physical Sciences (MPS), and the Directorate for Social, Behavioral & Economic Sciences (SBE)), and each directorate is composed of a number of divisions, organized according to scientific and technical discipline within each area. The NSF program directors within each division have responsibility for managing a technical portfolio in each scientific and technical subject area.

The NSF will periodically offer funding opportunities for specific research topics, which are determined through a strategic planning process organized to identify new and emerging research topics and areas.

These specific opportunities are announced by means of either a research solicitation (called a Dear Colleague Letter) or a request for proposals. Each division generally publishes their own solicitations and requests for proposals. The divisions will also accept unsolicited proposals on any topic consistent with their overall disciplines and interests. However, unsolicited proposals are accepted only during the "open window" period for each particular division. There are, in general, two windows, one in the fall, generally in the September/October time frame, and one in the spring, usually in February. However, some divisions offer only one window each year, which is usually in the fall. You'll need to check the NSF website and the Grants.gov website to stay current on the opportunities for proposal submission. All proposals submitted to the NSF are evaluated by means of a merit review procedure, and are evaluated by a panel of scientists, engineers, and experts selected based upon their expertise in the subject area of the panel on which they serve. The panel members are primarily derived from university faculty and scientists, but also may include scientific and technical experts from US government agencies, industry, and independent research centers and laboratories. For more information concerning NSF opportunities, policies, and procedures, you can explore the NSF website (http://www.nsf.gov). Along with the NSF website, research program opportunities are published on the Grants.gov website.

6.3.3 The US Department of Defense (DOD)

The US Department of Defense (DOD) has historically provided research support to academic scientists and engineers dating back to the founding of the nation. However, significant support for academic research primarily dates back to the nation's experience in the Second World War and the benefits to the military that were demonstrated by advances in technology. The nation's military forces are heavily dependent upon technological advantage that requires significant advances that are derived from research in science and engineering. As science and engineering advance, technological advantage requires ever-increasing levels of research to provide seed ideas and concepts, thereby providing the basis for the realization and development of systems that can provide the desired performance. The DOD is a mission agency and, as such, it provides support only for technical disciplines that are considered to be fundamental and supportive of the mission. In general, there are 12 technical areas that are appropriate for research support by DOD research offices. The 12 areas are tabulated here.

• *Physics*	• *Mechanics*
• *Chemistry*	• *Terrestrial Science*
• *Mathematics*	• *Ocean Science*
• *Computer Sciences*	• *Atmospheric and Space Sciences*
• *Electronics*	• *Biological Science*
• *Materials Science*	• *Cognitive and Neural Sciences*

These areas consist predominately of engineering disciplines, computer and information sciences, and physical sciences, including materials science and engineering. Mathematics research is also supported to a significant extent. The department also provides some, although in general limited, support in the social sciences, medical research, and life sciences. Research support in the latter disciplines is directed towards research consistent with the DOD mission, and this research does not significantly overlap or compete with research in these areas supported by the NIH or NSF. A significant strategic planning process is employed by the DOD, and agency priorities are continually being revised in accordance with emerging threats and scientific opportunities. Currently, seven priority areas for DOD-wide research and development have been identified. These seven priorities are Autonomy, Countering Weapons of Mass Destruction, Cyber Sciences, Data-to-data Decisions, Electronic Warfare, Engineered Resilient Systems, and Human Systems. These priority areas are subject to review and redefinition on a periodic basis. Each service and DOD agency also identifies their specific priorities and interest areas, and periodically they will identify specific focused research opportunities, sometimes identified as Grand Challenges (e.g., ONR's Basic Research Challenges, and AFOSR's Discovery Challenge Thrusts).

Although the DOD's budget for academic research in science and technology is relatively small compared with the budgets for the NIH and NSF, as indicated in Fig. 5.6, the DOD is, in fact, the major source of research funding in the disciplines in which they have priority interests, and therefore invest significant funds. For example, while the DOD provides only about 6% of the total funding for academic research in all disciplines, the department provides slightly under 30% of the research funding in engineering disciplines, around 20% of the research funding in computer science, and slightly under 20% of the research funding in mathematics. The DOD also provides significant funding for academic research in physical sciences and environmental sciences, providing slightly over 10% of the total funding for both disciplines. In certain disciplines, the DOD is the major source of research funding, providing almost 90% of the academic research funding for mechanical engineering, over 60% of the research funding for electrical engineering, and about a third of the research funding in metallurgy and materials science. The DOD's support for academic research derives primarily from the basic research (i.e., the 6.1) account, and the DOD provides about a half of their basic research funds for support of academic research. The department also provides about 15% of the applied research (6.2) funds, and slightly under 10% of the advanced development (6.3) funds for academic research. However, the majority of the applied research and advanced development (6.2 and 6.3) funds are provided to University Affiliated Research Centers (UARCs) and other specialized university research organizations that are structured to directly address and accommodate DOD security and other requirements. In general, these funds are not available to the wider academic research community.

The DOD is organized to interact and work with college and university researchers, and it hosts a variety of research offices that work extensively with the academic research community. Each of the three military services sponsors a research office that supports research programs within the academic community. The three service research offices are the Army Research Office (ARO), the Office of Naval Research (ONR), and the Air Force Office of Scientific Research (AFOSR). Collectively, the three offices are call the OXRs, where the X refers to the service and the O stands for Office and the R stands for Research. ARO is located in Durham, NC, and ONR and AFOSR are located in the Ballston area, in Arlington, VA, across the Potomac River from Washington, DC. In general, ARO and AFOSR manage basic research (6.1) funds, while ONR manages the science and technology and advanced development (6.1, 6.2, and 6.3) funds for the Navy. Academic research is also supported by the Defense Advanced Research Projects Agency (DARPA), the Defense Threat Reduction Agency (DTRA), and the Defense Medical Research and Development Program (DMRDP), and sometimes other DOD organizations. The main mechanism for DOD basic research funding to academic organizations is a grant, primarily through one of the OXRs, DARPA, or DTRA. However, the DOD also supports a significant number of applied research and advanced development (6.2 and 6.3) projects, primarily through the services in-house research laboratories (i.e., the Army Research Laboratory (ARL), the Naval Research Laboratory (NRL), and the Air Force Research Laboratory (AFRL)), but also through DARPA, DTRA, and the Missile Defense Agency (MDA). Sponsored research from these organizations will generally be provided in the form of a contract, and quite often will involve partnerships with industrial collaborators and colleagues.

The best places to begin your exploration of DOD agency and office research interests and priorities and the opportunities for research support are the various office websites. Since the DOD is large and complex, there are many locations to explore. The primary websites are as follows.

Army Research Office:	http://www.arl.army.mil/aro
Air Force Office of Scientific Research:	http://www.afosr.af.mil
Office of Naval Research:	http://www.onr.navy.mil
Defense Advanced Research Projects Agency:	http://www.darpa.mil
Defense Threat Reduction Agency:	http://www.dtra.mil
Missile Defense Agency:	http://www.mda.mil
Defense Medical R&D Program:	http://www.dmrdp.dhhq .health.mil

6.3.4 The Department of Energy (DOE)

The Department of Energy (DOE) supports basic research and provides research grants to academic institutions and industries, primarily through the DOE Office of Basic Energy Sciences (BES). About 40% of the BES budget provided to external organizations is allocated to academic researchers. As a mission agency, the DOE, through BES, is interested in sponsoring research directed towards building understanding, and establishing foundations associated with energy, the environment, and national security. The BES program is a major sponsor of basic research in the natural sciences, primarily condensed matter and materials physics, chemistry, geosciences, and aspects of physical biosciences. The DOE BES focuses research support into disciplines that are directed towards advancing discoveries in new materials, new chemical processes, research related to all areas that involve energy resources, including production, conversion, waste mitigation, transmission, storage, and efficiency. BES is interested in basic and fundamental research directed towards understanding, predicting, and ultimately controlling matter and energy at the electronic, atomic, and molecular levels in an effort to establish the foundations for new energy technologies. The BES program also provides support for large-scale, scientific user facilities at various locations. These facilities house instrumentation and experimental measurement equipment for the purposes of imaging, materials analyses and characterization, and understanding chemical transformation. Materials that can be characterized include a wide range of substances, extending from metals, metal alloys, and ceramics, to biological samples. Characterization instrumentation permits research on both the microscopic and macroscopic levels to be conducted. Nanoscience and nanotechnology research are also focus areas for BES research, and significant support is provided for projects in these areas. The overall goal of research sponsored and supported by BES is to provide a base of knowledge that will permit the understanding and establishment of the scientific basis necessary to design technologies that can adapt to the natural world and secure a sustainable energy future for the nation and the world.

In order to maintain an energy focus to the BES basic research program, BES establishes distinct Core Research Activities (CRAs), which define the scientific disciplines that address the scientific and engineering base for the various energy technologies. The CRAs are structured as scientific disciplines, rather than technology areas, and are designed to align with the BES organization.

The Advanced Research Projects Agency-Energy (ARPA-E) provides funding for applied research and development projects, but does not provide support for basic research projects. Nonetheless, many of the applied research projects are appropriate for academic research and their funding opportunity announcements should be monitored for possible response.

The primary websites for DOE sponsored research opportunities are as follows.

http://science.energy.gov/bes/
http://arpa-e.energy.gov

6.3.5 The National Aeronautical and Space Administration (NASA)

NASA provides support for a wide range of science and engineering basic and applied research topic areas associated with the NASA mission, which is: "To pioneer the future in space exploration, scientific discovery, and aeronautics research." NASA directs their external research support to projects that contribute to its space or airborne assets, which could include making use of the assets, or research that is directed towards making use of the data derived from the assets. The NASA research program is organized around four mission-oriented directorates, each with their own research interests. The four directorates are: (1) the Aeronautics Research Directorate (ARMD), (2) the Human Exploration and Operations Directorate (HEO), (3) the Science Mission Directorate (SMD), and (4) the Space Technology Mission Directorate (STMD). ARMD is interested in research directed towards determining solutions to challenges that exist in the nation's air transportation system, including air traffic congestion, safety, and environmental impacts. Research interests include topics focused upon green aviation technologies that will enable fuel-efficient flight operations, along with reduced emissions and noise. Research directed towards new aircraft technologies, including systems-level research on the integration of new operations concepts, is supported under the ARMD program. The HEO program is focused upon research and development (R&D) activities directed towards advancing human and robotic space exploration. Directorate interests include human exploration in and beyond low-Earth orbit. Exploration activities beyond low-Earth orbit cover a range of technologies associated with commercial space transportation, exploration systems development, human space flight capabilities, advanced exploration systems, and space life sciences and applications. The HEO Directorate is also responsible for issues associated with launch services, space transportation, and space communications in support of both human and robotic exploration. The Science Mission Directorate (SMD) has interests in research programs directed towards providing the science basis for a mission, defining technologies and techniques necessary to actually execute a mission, establishing technologies and techniques for gathering, calibrating, validating, and analyzing data during a mission, and analyzing and archiving data gathered during missions for later analysis. The overall goal of the SMD programs is to make use of the vantage point of space and provide the science community with the platform and tools to investigate and increase understanding of our planet, other planets, solar system bodies, the interplanetary environment, the sun and its effects upon the solar system, and the greater universe. The STMD programs are directed towards crosscutting, pioneering, and new technologies and capabilities needed to achieve current and future missions. A wide variety of disciplines are involved in this effort, with the goal of maturing and advancing technology required for space exploration missions. Many of the technology advances find use in other government agencies and commercial space activities.

NASA research opportunities for university-based research are announced periodically by means of NASA Research Announcements (NRAs). The NRAs are published on the NASA website, NSPIRES, which lists the NRAs, as well as other information associated with proposal submission. The NRAs are, of course, also listed on the Grants.gov website.

The primary websites for NASA sponsored research are as follows.

http://nspires.nasaprs.com/external/

http://science.nasa.gov/researchers/sara/how-to-guide/

The NSPIRES website contains links to each of the mission directorate's websites, where more information concerning each directorate can be found. The NASA Science for Researchers website (the second website listed) contains useful information concerning proposal submission requirements and procedures.

6.4 How Do I Identify and Make Contact with a Program Manager or Program Director Appropriate for my Research Interests?

In this section we'll address some common questions often expressed by young and new faculty members recently recruited to an academic faculty position.

Question 1: I'm a new PhD graduate in electrical engineering, and I've just accepted a faculty position as an Assistant Professor. My department head has informed me that he expects me to write proposals to obtain research funding to support my research and my students (that I need to recruit). He's assigned me a light teaching load for the first year and provided me with some start-up funds. However, the amount of the start-up funds is limited and not sufficient to fully fund my research activities. I really need to get a research grant in place as soon as possible. Where do I start?

Answer 1: Your question is very appropriate, and you are in a position similar to that of a large number of new faculty members in the initial stages of their academic careers. Success in your new career depends very much upon how you approach this process. The answer has several dimensions. First, you need to determine your research topic.

This may appear simple, and your first impulse is, most likely, to continue your thesis research since this is the area in which you have the most experience and the most complete knowledge base. You've most likely published some of your work in peer-reviewed technical publications, and may have given papers on your research at professional meetings and conferences. You may have even received significant professional recognition for your research, either from your university, or from professional societies, funding agencies, etc. It may be very tempting to continue on the same work. However, upon reflection, your thesis research may not be the best work upon which you wish to build your career. For example, your research thesis topic may have been simply assigned to you by your mentor, and you may actually not have a major interest in continuing to pursue the topic and would prefer to work in an area you find more interesting. Also, you may sense or be aware that your thesis research topic, which was once a very good and intriguing idea, has not experienced significant and positive progress, and the topic is losing interest within the professional scientific and engineering community. Over the course of my career I have witnessed numerous scientific and engineering ideas and approaches emerge, usually accompanied with much fanfare, enthusiasm, and significant research effort and financial support, only to find that the early optimism and enthusiasm were significantly over-estimated and overstated, and, after several years of intense research, positive research progress was limited, and interest subsequently waned.

Many of these less-than-successful research topics were, for a while, the dominant topic within certain disciplines, complete with "special sessions" at technical conferences, research panel discussions, meetings, journal articles, technical books published, etc., and, of course, supported with significant financial support from funding agencies. However, without positive progress being achieved, interests shift to other topics. When this occurs, funding agencies shift their support to other topics and alternative approaches. Also, when significant progress is not achieved and interest wanes, the researchers who had focused much of their effort on these research topics do not simply go away with the topic, but rather they refocus their research interests to alternative approaches, or to entirely new research areas. Researchers almost always will follow the money, and shift their research activities to areas that are experiencing a growth in research funding. In one respect, these researchers are in a better position for their next research challenge since all research is a learning experience, and the knowledge gained in the process of researching a topic almost always proves of value in subsequent research efforts.

Another potential reason that your thesis research may not be the best place to concentrate your research activities relates to the level of development of your thesis research topic. Research is a very dynamic process, with topics continually emerging, undergoing intense research effort and, if the research is successful, the topic may mature and transition to advanced development and practical applications. If your thesis research is in this category, and undergoing a transition from basic to applied research, you should consider your position in the field. That is, you need to assess the status of your research and seriously consider what you can contribute to future advances in the field. The correct decision for you may be difficult to determine, but will be related to how you assess your status in the field and how you view the future. For example, in your new position as an Assistant Professor, will you have the research facilities necessary to continue to make advances in the area? If the research area has demonstrated significant progress and has developed to the point where practical applications are beginning to emerge, it is likely that industrial or commercial interests have developed. There may even be new start-up companies appearing with interests and products based upon the topic related to your thesis research.

This situation, which is actually fairly common, can be a two-edged sword with both positive and negative aspects. A negative aspect is that you may find yourself in an area with a growing number of contributors and locations performing related work. A good idea tends to spread rapidly, with many other scientists and engineers quickly shifting their efforts to related work. You may find yourself in competition with scientists and engineers working in much better equipped industrial laboratories. In this case you may find your knowledge base quickly aging as the research in the field rapidly progresses and expands. A positive aspect of the process is that your thesis research has, most likely, provided you with state-of-the-art and detailed knowledge that positions you as an expert in the field. You may have been one of the first researchers to make contributions in the field. Your knowledge of the subject could make you a very attractive collaborator with the industrial scientists and engineers working on the subject. Such collaboration often serves as an excellent source of funding for your continued research efforts, as well as providing exposure and access to well-equipped industrial laboratories and facilities that may not be readily available to you at your university location. You may find that your ability to advance the field is greatly enhanced by the industrial augmentation. I have worked extensively with industrial researchers throughout my career, and I have found that my research was significantly enhanced by the collaboration. Not only have I received significant financial support from this type of collaboration but, more importantly, I was exposed to many real-world problems encountered in the field that served as the focus of my research and kept the work at the state-of-the-art. The industrial collaboration provided me with research data that would have been either difficult to gather or not available from the university laboratory resources at my disposal. In my experience, industrial scientists and engineers value collaborating with academic researchers, both for the research advances, and also for access to students that often result in future employees for their organization.

Recognition of the status of your research can be very difficult to determine and requires maturity on your part, a maturity that will improve as you continue in your career advancement. Nonetheless, you need to question the status of your research topic and determine if you wish to continue on your current research course, or if you need to build upon what you have already achieved, but move in new directions. Most faculty members find that their research performed as a graduate student prepared them for a career in research, but that the subject of their subsequent research differed from their thesis research, and often by significant deviations. How you assess your situation is very important since it will dictate the topic of your research proposal, as well as focus your efforts upon the organization you wish to pursue for financial support.

Question 2: OK, I've evaluated the area in which I perform research, I've spent a fair amount of time assessing my research and the results I obtained, as well as my status in the field, and I've determined the research topic I would like to pursue. After reviewing my own research, as well as that reported by others, I have identified a very promising topic, which is related to, but takes a new direction from my thesis research. I have a great idea for what I think is a very good research proposal, and I even have some results and data that are supportive of my idea. I've started to survey potential funding agencies, and I've searched the agency websites, found one with research grant programs that are in my interest area, and even identified a program manager who supports research similar to my own. I've even tried to contact him, both by email and by phone, with no success, and I haven't been able to get a response. What am I doing wrong? And where do I go from here?

Answer 2: You aren't doing anything wrong. In fact, you're doing exactly what you should be doing, and you are following a good approach. Quite often the program manager or program director will immediately respond to your attempts to contact them, but other times you may have delays in getting them to respond, and sometimes they won't respond at all. You need to keep in mind that program managers and program directors are very popular people with those wishing to acquire research support, as well as very busy people with their jobs and professional commitments and, at times, can be difficult to contact and get to respond. Basically, you need to develop stamina and perseverance, and keep trying. Generally, their failure to immediately respond to you, when this occurs, is more based upon their work commitments than any desire to not communicate with you. For the mission agencies in particular, personal contact with program managers is extremely important for success in obtaining grant funding. Keep in mind that their job requires them to stay at the forefront of the state-of-the-art in their technical discipline, and if you have interesting results and a new approach to a specific problem, and particularly when your approach is supported with experimental data and/or other supporting evidence, they most definitely want to learn what you are doing. Also, it is useful to recognize that program managers are always on the lookout for new and promising scientific and technical talent who can bring a new and fresh perspective and approach to challenging problems. Also, funding agencies have a variety of special awards for new and young investigators. These awards are separate from their regular core program and they receive credit in their performance evaluations when one of their researchers is recognized with one of these awards. Since you are potentially a candidate for one of these awards, they have a natural desire and willingness to meet you and learn your ideas and approaches to challenging problems.

However, the impetus for establishing the first contact with a program manager is your responsibility, and until you actually make contact with them and establish a dialogue with them, they won't know what you have to offer. For this reason, you need to be very persistent in your efforts to communicate with them. If sending them an email doesn't result in a reply within a reasonable length of time, which could be on the order of a few days or a week, you should try to contact them directly by telephone. You can either get their phone number from the agency website, or call the office in which they work. The agency telephone operator or division secretary will either connect you with them, if they happen to be in their office, or will take a message and make sure it is delivered to them. When you make contact on the telephone with them, you should keep your discussion very brief and for the purpose of establishing personal contact. It is highly unlikely that a phone call alone, especially an initial phone call, will prove sufficient to gain their interest to the degree that they will welcome a proposal. Rather, your efforts should be directed towards providing a very brief overview of the research you wish to pursue and to determine their interests. Your main objective is to find a basis of common interests and to learn how you can fit into their program. This will not be possible on a first, brief telephone call, but you should be able to confirm that you have identified the correct person for your interests.

If not, you should question the program manager concerning which program manager is a better match for your research interests.

The funds that any program manager has available for new grants are very limited and they will only provide financial support for research that is directed towards problems in which they have interests. The good news for you is that the program managers are always receptive to new approaches that show potential for making advances in the areas that support their goals and objectives. Your job is to learn what goals and objectives are being pursued by the program manager. Therefore, the initial telephone call should be directed towards briefly introducing yourself and giving a brief (short) overview of your research, with emphasis upon your new and novel approach, and starting a discussion with them on their program and the research that is currently being supported. You want to learn the scientific and technical interests of the program manager, and attempt to learn the overall goals of their research interests and support. If your initial telephone call is productive, there may be opportunity for follow-up calls, or preferably, a personal visit. In fact, as a priority, the next step for you is to attempt to make an appointment with them for a personal visit. This step is extremely important as a means to establishing communication with program managers in mission agencies. This will require some travel on your part, along with the associated travel expenses. Hopefully, your department will provide the necessary travel support, or include travel support for funding agency visits in your start-up package, as discussed in Chapter 3.

Once you schedule a personal visit with the program manager you will, most likely, be limited to a half hour, or so. This meeting will be very important so you'll need to be very organized and effective with the limited time you'll most likely have. You should prepare a very brief overview of your research, preferably using a very limited number of viewgraphs or slides. In my experience, faculty members generally have slides prepared using PowerPoint on a laptop computer. The presentation does not need to be formal or rehearsed. In fact, it's better not to be too formal, since a main objective is to get the program manager engaged in your work and to pique their interest. You want them to ask questions. Since time is limited you should focus upon the main topic of your research, what approach you are following, what results you have obtained to date, and the direction you wish to pursue. All of this should be presented in no more than four or five slides. This generally will take some planning effort on your part since it's more difficult to give a short presentation than it is to give a longer and more detailed presentation. New and young faculty members, in particular, often have difficulty in giving short presentations since they want to present details. However, for the purpose of building communication with program managers, the focus should be on general approach and trends, and overall goals. Details will detract from what you are presenting, since it opens the discussion to issues that require more time, generally more than you will have at the meeting. Scientific and technical details are better left to future discussions, hopefully in the process of securing a research grant. Of course, if the program manager requests detailed information, you should be prepared to offer explanations, details, published results, etc. If you have published papers on the subject or closely related subjects, you should bring copies of the publication that you can leave with the program manager. Also, during the meeting and after the technical discussion, you should inquire about funding opportunities. You'll generally receive a general and noncommittal response, but you'll have broken the ice and planted some ideas for future discussion. Following the meeting, you should wait a few days and then follow up with the program manager and seek to determine the interest in your research, whether it is consistent with the program manager's overall goals, and how you might modify what you are pursuing so that it is more consistent with the program manager's objectives. This process may take some time, but if done properly, can lead to good success.

Question 3: Are there other techniques or methods for making contact with a program manager or program director besides telephone calls and email? What do I do if I can't reach them by telephone and if they don't answer my email messages to them? They don't yet know me, so how can I meet them?

Answer 3: This is another very good question, and this situation does occur, sometimes (and unfortunately) much more often than it should.

The answer is that it is very possible to make contact with the program manager to whom you would like to communicate. However, the process may require some time, and much more effort than a simple telephone call or email.

You've already determined that the program manager has research interests similar to your own. You can use this as an advantage. First, you'll need to travel, which, as previously discussed, will require support from your home institution. Of course, you can schedule visits to the program manager's office. But often it's more effective and convenient to meet program managers at technical conferences, meetings, and reviews. You're going to want to attend many of these meetings since it's one of the best methods to stay current in your technical specialty. It's at these meetings where the most current state-of-the-art results and developments are presented and discussed. Program managers will also attend these meetings, and for the same reason. Therefore, there is a good chance that you'll have the opportunity to meet an appropriate program manager by introducing yourself during a coffee break or other pause in the presentations. You'll often find small groups of meeting attendees talking in the halls, including researchers and program managers. Meeting attendees always wear name badges with their name and organization clearly displayed, making identification an easy process. In fact, meeting organizers always schedule breaks in the formal presentations explicitly for the purpose of attendees having an opportunity to meet each other and discuss mutual interests. In my own experience I've met many people, both researchers and program managers, by exactly this process. Over time you'll find your circle of contacts increases, and you'll meet many people that will become colleagues with whom you'll maintain communication, sometimes extending over many years. Networking of this type is common and widespread and is a primary method both of making new contacts, as well as maintaining previous contacts. Attendance at appropriate technical and professional meetings will be an important part of your professional development.

Participation in appropriate professional and technical conferences, meetings, and reviews as an attendee is the first step, but you'll also want to attend these meetings to present the results of your work. As your research progresses, you should make every attempt to submit your work for review and inclusion in the meeting proceedings. This permits you the opportunity to stand before an audience of your technical peers, including program managers, and present the results of your work. If your work is closely related to the interests of the program manager, it is likely that they'll be in the audience. After the meeting you can also take the opportunity to send a copy of your paper directly to the program manager. As a program manager I've received many copies of papers and research reports from numerous researchers, sometimes on a fairly regular basis. There is a fairly large group of researchers who regularly send their papers and research reports to a distribution list of program managers. This process can be effective because even if the program manager doesn't read everything sent to him or her, he will at least read the title and names on the work, and if the title is of interest, he will read the abstract, and if that is interesting, he'll read the entire work. Also, simply receiving the paper and reading the title and author will get the author's name in the program manager's mind. Your subsequent attempts to communicate with the program manager will, likely, be more effective.

While participation in professional conferences, meetings, and program reviews, both as a participant and presenter, is extremely important, you should also volunteer to participate in the planning and organization of these meetings. Becoming involved in these activities is actually quite easy since the planning, organization, and management of these meetings requires a fair amount of effort, and the majority of the effort is performed by volunteers. The organizers are always looking for new volunteers. The conference organizers and their positions are always listed on the meeting announcement and schedule, both the print version and the electronic website version. You should contact them, identify yourself and your credentials, and volunteer your services.

Question 4: I've had a personal meeting with a program manager, I gave them a short presentation on my research, and had a very interesting discussion on future directions. The program manager seemed very interested in my work, which I consider to be very encouraging. However, he then gave me the bad news, and indicated that his core research budget is very limited and he is currently over-extended. He stated that it would be very difficult to fund me at the present time. But then he asked me to submit a "white paper," and mentioned that there was a possibility for "year-end money." What's a "white paper" and what is "year-end money"?

Answer 4: Congratulations! You've made excellent progress. You appear to have made a good impression and presented ideas of interest to the program manager. You are on your way to establishing a good basis for future discussions and communication with the program manager. The program manager's response to you is very encouraging, and you need to focus upon responding with the requested information as a high-priority action on your part. First, let me explain the concept of a "white paper," and then I'll address "year-end money."

6.4.1 White Papers

The program manager has expressed interest in your research and has asked for you to prepare and send him a "white paper." The white paper is a document that gives you an opportunity to provide the program manager more detailed information on your research in a concise and simple way. The white paper does not have a formal format and is fairly easy to prepare. However, since it is a written document, it serves as an excellent follow-up to discussions, and it provides an effective means to keep your name and ideas readily available to the program manager. White paper submissions are becoming more popular and common for funding agencies since they offer enhanced flexibility in research program planning. It is also becoming more common for white paper submission to be requested in the funding agency BAAs and research program opportunity announcements, solicitations, etc. Use of white papers saves much work by both the proposer and the program manager since they are generally brief and don't require significant time to either write or to evaluate. In addition, the white paper is not an official proposal and does not require formal action by your home institution or the government agency.

It's simply a document that outlines, in a brief manner, usually five pages or less, your thoughts, ideas, and proposed research. The white paper does not require any official action by the government agency, and after submission you may or may not receive a response from the program manager, sometimes for extended periods of time, and sometimes not at all. However, the white paper may provide information that is of significant interest to the program manager, and this information may permit the program manager to plan his research program, with the positive result that he contacts you with a request for a formal proposal. You should carefully prepare the white paper and send it, as requested, to the program manager. I emphasize that the white paper should be brief, and certainly fewer than five pages. One- or two-page white papers are generally adequate.

There is a more formal white paper procedure often employed by funding agencies. When white papers are requested in published BAAs or research opportunity announcements, particularly for specific topics with defined funding, the white paper is often used to determine agency interest in your research and if it is appropriate for the announced research opportunity. Under these conditions your white paper will be evaluated and you will receive a response from the agency, generally by a defined date. The response will either encourage or discourage a formal proposal submission. If the response discourages a formal proposal submission, you may receive or you may request a debriefing explaining the reasons for the agency response. The white paper procedure is intended to discourage the time and effort required to prepare the submission of proposals that have little chance for success, and to limit the grant competition to the proposals that best address the research program objectives. This type of formal white paper process is often used for larger research opportunity programs, such as research centers, where significant time and effort is required to both prepare and evaluate long and detailed proposals, or for specific research opportunities addressing specific topics.

The BAA or research opportunity announcement may request that a white paper have a defined format, and request specific information be addressed that is generally explained in the BAA or research opportunity announcement. When specific information is requested, care should be taken to address all questions and issues discussed in the BAA and include the appropriate responses to all information requested. The responses and the details of how this information is addressed by the principal investigator is considered by the evaluators, and the responses included in the white paper are primary factors in making the decision regarding whether a full proposal will be encouraged or discouraged. Typically, the BAA or solicitation will define the length of the white paper, and will request a white paper on the order of 10 to 15 pages in length, and sometimes no longer than two to five pages. No matter the length of the white paper, it should be carefully prepared with attention to the details requested in the BAA or announcement. When white papers are specifically requested in the BAA or announcement, they may need to be submitted through the official university sponsored research office, depending upon the grant proposal submission policies of your university. However, the white paper is still not an official legal document. If you receive a discouragement, the discouragement is not legally binding, and a proposal can still be submitted and will be considered and evaluated by the funding agency. However, a proposal submitted after receiving a discouragement should be significantly modified from the white paper and should address the weaknesses and inadequacies that resulted in the white paper receiving a discouragement decision.

6.4.2 Year-End Money

The concept of "year-end money" refers to a mechanism that program managers, particularly those in DOD agencies and offices, often use to initiate new grants in which they have interest, but don't have sufficient funds available in their core program. The term "core" refers to the funds that they are allocated to support their main research interests, which are determined in the office or agency budget by means of the agency strategic planning process. The "year-end money" mechanism provides them an opportunity to provide "seed" type funding for new and novel approaches that they feel aren't sufficiently defined or developed for a regular grant award. The concept of "year-end money" refers to the internal budget process for the funding agency and how they provide for incremental funding requirements. United States government agencies have funding only as a result of Congress passing an appropriations bill for each agency, and the President signing the bill to turn it into a law. Once the appropriations law is passed, the agency is then authorized to execute their budgeted funds. The budgeting process is performed on an annual basis, and the appropriations law provides funds to each agency on a fiscal year basis, which for the US government extends from October 1 of each year, to the end of September of the following year. The typical duration for a standard research grant is generally three years, although the time period will vary somewhat from agency to agency. However, due to the annual US government budgeting process, the funding agency program managers will have funds available only for the fiscal year in which an appropriations law exists, and a new law is required to provide the funds to support subsequent years in the multi-year grant period. This, of course, means that additional year funding for the grant will be available only if a new appropriations law is passed. Program managers must, therefore, be good financial managers and provide the increments to the various research grants they support on a timely basis.

Since most new awards will initiate on a date that does not correspond to the start of the fiscal year, and since some awards that are terminating will end on a date during the fiscal year, the amount of funds available does not necessarily match the amount of grant funds commitments that exist. Program managers will provide the funding increments to the grants supported in their program, according to the budget schedule defined in the proposal. These increments are provided on a priority basis, and any remaining funds are used to provide the funds for new grants that are being initiated. This process proceeds throughout the fiscal year, in response to the available funding. It is not uncommon as the end of a fiscal year approaches for a program manager to have a limited amount of funding available that has not been committed. Since the remaining funds need to be spent before the end of the fiscal year, the opportunity exists to provide funds to start a new grant. This is the source of the "year-end funds."

Generally, when a program manager decides to provide some of the remaining funds to a new principal investigator using the remaining year-end funds, the amount of funding that will be provided will be limited, and a grant for only a year or less in length will be awarded. The restriction to the one-year or less grant duration does not increase the program manager's commitment to providing future funds, and does not result in an over-extension of their budget. However, the one-year period permits the program manager to provide some "seed" funds in order for the new principal investigator to work on the research idea and further develop the concept or idea. The additional information can, if successful results are obtained, be used to support the preparation of a more complete proposal that can be submitted for consideration as a regular grant. The use of year-end funds is particularly attractive for program managers to provide seed funds to new principal investigators or to those who are changing research directions. Program managers make use of this mechanism only at the end of the fiscal year and, of course, only if remaining funds are available. Nonetheless, the mechanism is an attractive option to initiate new projects, and particularly those that are considered high risk, but potentially high payoff, in terms of the approach. Many program managers will use the year-end funds mechanism to support young and new principal investigators for short periods for the purpose of giving them an opportunity to demonstrate their research potential, and in fact, many of these year-end funds grants develop into regular grants in subsequent years.

Question 5: I submitted a white paper to the program manager. He read it and contacted me with a very encouraging response, and asked me to submit a formal proposal. However, before preparing and submitting the proposal, I'd like to know a little concerning how my proposal will be reviewed. I think this information will help me prepare the proposal. Can you provide a little background concerning proposal review procedures and considerations? Also, is there a difference between agencies relative to what they want to see in a proposal? I'd appreciate any comments you can offer.

Answer 5: I'll address, in detail, the information program managers expect to see included in your proposal in Chapter 7. The information will vary somewhat, depending upon the funding agency, and there are requirements that need to be addressed. However, before addressing the specific issues associated with preparing a proposal, it's useful to understand some general issues regarding how your proposal will be evaluated and reviewed. The process does differ by funding agency and it's useful to understand the differences before preparing your proposal. Also, the differences in the mechanisms used by various government funding agencies to provide grant funding are useful to understand. These two issues are discussed in the following sections.

6.5 Proposal Evaluation Considerations

United States government agencies, in general, base their proposal evaluations upon "peer review," which means that each research proposal they receive will be evaluated and reviewed by PhD-level scientists or engineers, or those with equivalent experience, with expertise in the technical area of the proposal. The peer reviewers will have expertise in the subject of your proposal, although they may not necessarily be active researchers. However, they will know the subject and be familiar with current research trends and developments in the area. Essentially all US government agencies manage and operate their research offices under the peer review principle, although for certain mission agencies, the actual peer reviewer may be the program manager who receives the proposal. Three to five experts, normally, evaluate each proposal. The experts will consist of scientists and engineers from within US government organizations, universities, and industrial organizations. Generally, for mission agencies the proposals are sent to the reviewers for evaluation, and the reviewers will perform the review at their home institutions. They will submit the reviews to the program manager who is managing the process. Many funding agencies have web-based systems for proposal review and evaluation. The program manager in the mission agencies, who as a PhD-level scientist is considered a peer, may be one of the reviewers, and will almost certainly read your proposal whether or not they actually perform a formal review. The reviewers do not normally meet as a group. The program manager will read the evaluators' comments, consider all reviews, and then make the decision regarding the acceptance or declination of the proposal. In this regard, the program manager has a very significant role.

For certain research program opportunities that are published as a specific Call for Proposals, the process is slightly different. In this case the proposals are received in response to the published solicitation. The proposals will be collected and organized according to the program definition, and organized by topics. The proposals will be evaluated by panels of experts that can consist of experts from US government organizations, universities, and industry. The US government program managers associated with the particular solicitation will then meet and decide which proposals are selected for acceptance.

The point here is that the program managers in mission agencies are all trying to accomplish a well-defined scientific or engineering advance. In order to accomplish their goal, they need a variety of scientific and engineering "building blocks" that all fit together into a complex web of interrelated scientific and engineering disciplines. The program managers know what areas and subjects they want to acquire for their programs. They also know who is doing what and at which institutions, both academic and industrial, and who is performing the best work in which they have interest. They generally have already provided, and are currently providing, funding support for various research topics and subjects to a variety of researchers, and consider these researchers to be fundamental to their interests. Each of the researchers in their program are essentially serving as "building blocks" for their overall goals and are helping to realize what the program managers are attempting to accomplish. The overall program strategy can consist of a wide variety of end goals, ranging from a scientific advance in a given theory, material, device or component, to realization of an advanced system. The overall goal can be narrow in scope or very large, involving multiple program managers in multiple offices and even across agencies.

The important point for you to recognize is, if you wish to obtain funding from any of the mission agencies, you are going to need to learn the specific interests of the program managers, what they need to accomplish their overall goal, and how you can "fit" into their program. You'll need to identify the program manager most appropriate for your research. Your research area may be an excellent fit to the program manager you are contacting. However, after communicating with the program manager you may find that you need to "tune" and modify your research objectives slightly and refocus your research to areas of interest to the program manager. Basically, you want to become one of the "building blocks" in the program manager's research program. Once this is accomplished, you'll find that the program manager will, most likely, become a strong supporter of your research, particularly if it's successful, and you'll be well on your way to establishing an effective and productive enduring collaboration. Once you become an integral part of a program manager's program you'll find that they may become a supporter of your research work and will make every attempt to keep you funded to work on topics that are important to his or her objectives. Program managers have a variety of options to obtain funding from within various offices both within and outside their home agency. It's common practice for program managers to seek research funds from other internal government sources to support and augment their core activities. Most program managers actually manage a portfolio much larger than their core program, supported with funds they solicit and receive from their colleagues in other US government offices, laboratories, and field operations.

The program directors at the NSF have similar objectives, but function in a slightly different manner. The divisions in the various directorates at the NSF are organized according to technical area, with a program director in charge of each specific area. The technical areas are often fairly diverse and can include certain topics that only generally relate to each other. The reason for this is that the NSF accepts proposals from the entire academic community on virtually any scientific, technical, or educational research subject, as long as it can be related to the interest areas associated with each division. This opens the proposal topics to a wide range of scientific and technical subjects. Once the proposals are received, the program directors will organize them into groups of similar and, hopefully, related topics. If a program director is assigned a proposal that doesn't fit well into the group of proposals they are managing, they will trade or transfer the proposal to another program director who will accept the proposal for review if the topic is appropriate for the subjects in their technical portfolio. Proposals that are not accepted by a program director will, most likely, have a low probability for success and generally will be declined. For this reason, it is important to make sure your proposal is submitted to a program with interests and goals consistent with that program.

At the NSF proposals are organized into groups based upon similar topic and subject matter, and then assigned to a panel for review and evaluation. The review panels consist of experts from organizations outside the NSF, generally universities, but also from industry, and other government organizations. The expert review panel members are selected by the program director from a database of qualified reviewers, and sometimes from their professional colleagues and contacts. The review process is a true "peer review," since the reviewers are all research level scientists and engineers. The role of the program director is primarily administrative, and focused upon the management of the review panel, and to ensure that each proposal is fairly and equitably reviewed and evaluated. However, the program director generally does not review and evaluate the proposal personally. Nor do they generally enter, in any significant detail, into the proposal evaluation discussions within the panel. This, of course, varies with the program directors, and some will actively participate in the discussions, although their main function is to lead the review and evaluation discussions. The panel members will discuss the details of the research and the research performance plan contained in the proposal, and will rank the proposals according to metrics such as overall quality, importance of the research topic, outreach plans, adequacy of budget, etc., and place the proposals into categories of: (1) recommend for acceptance and funding; (2) recommend for acceptance if funding is available; and (3) decline and do not fund. The program director will then determine the number of proposals and which ones will be recommended for acceptance and funding. The actual decision on which proposals are funded is the responsibility of the division director, who is responsible for managing the division budget. The division director will either concur with the recommendation of the program director, or return the proposal to the program director for modification of the recommendation, or declination.

The main difference between the goals of the mission agency program managers and the NSF program directors is that the NSF program directors are simply looking for the best quality proposals to support. They are not necessarily looking to construct a well-defined and coordinated program focused upon a specific end result goal. They do not necessarily consider individual research programs as "building blocks" that relate and coordinate in an orchestrated manner. Rather, they attempt to find and support the most novel approaches to basic research in areas of science, engineering, and education research in their discipline. The mission agency program managers are focused upon achieving advances in a specific topic or area and are interested in constructing an overall research program that involves research advances from multiple performers. For this reason, they are not necessarily interested in funding the highest-quality proposal, particularly if the proposal subject is not in an area that is important to their overall goals and objectives. All program managers and program directors, of course, seek to identify and support the highest-quality research work and to accept the highest-quality proposals within their specialty areas.

6.6 Research Grants Basics: Standard Grants, Follow-Up Grants, and Incremental Funding

The basic research grant is termed a "Standard Grant," and is funded for a specific performance period of time, generally three years, although the actual performance period can vary. Once you receive your first funded grant from a program manager and perform successfully and produce meaningful results, you'll be in a good position to receive a renewal for your grant, or a follow-on grant for additional research. Mission agency grants, in contrast to those from the NSF, can be renewed for an additional period of time, and multiple extensions are also possible. Once successful relationships with program managers have been established, you may work with the same program manager over a sustained period of time that can extend over many grants and many years. For example, I've known researchers who have been supported by the same program manager with grant funding over a period extending over a decade, or more. Once the program manager gains confidence in you as a researcher, and has learned to value your results and how your work fits into their overall program objectives, they have many means at their disposal to secure funding for your research. These follow-on grants, by the way, do not necessarily need to go through the intense competition process required for responses to specific calls for proposals, etc. That is, research grants provided from the program manager's "core" program go through a proposal review and evaluation procedure, but the proposal is not necessarily in extensive competition with other proposals. The program manager can fund your proposal based upon the subject matter, which will be covered under the general BAA issued for their agency. In their proposal evaluation they will indicate that your proposal was selected from the number of proposals that were received under the BAA, but in actuality there may be a very small and limited number of proposals that were in competition for the award due to inappropriate subject matter or other reasons. There are also no time limitations on when the proposal is submitted, and you can submit at any time during the year. However, you need to be aware that most program managers have a limited amount of funding available to allocate to their core program. The actual amount of core funding varies from agency to agency, but is always limited and defined by agency budget details.

Also, most mission agencies will fund research grants on what is called an "incremental" funding basis. That is, grant funds are always provided for a determined time period, which is generally three years for standard grants, with an increment of the total amount provided on a yearly basis, determined by the research performance requirements. These requirements are clearly defined in the grant proposal budget.

The "incremental" status indicates that the grant funding will be provided on a yearly basis, based upon the US government fiscal year, which runs from October 1, each year, through the end of September in the following year. In order to understand the incremental funding process, and how it can affect your research program, it is useful for me to briefly digress and review the US government budgeting procedure for the various agencies and discuss how the research budgets are determined. The intent here is not to explain the government budgeting process in detail, but rather to give you a brief overview of the process and how it relates to your efforts to obtain research funding in your effort to build your research program.

6.6.1 The US Government Research Budget Process

United States government research agency budgets are determined through a complex budgeting process that involves their agency management, the national Administration, and Congress. The entire budget process is managed by the White House Office of Management and Budget (OMB), which works with the various government agencies, as well as Congress, throughout the year to define the agency budget requests that will be submitted to Congress. Basically, the Administration defines and requests budget funding, but only Congress can appropriate funds. The Administration establishes agency budget priorities depending upon a variety of issues, and works with the agencies to determine budget requests. Input from all agencies are gathered and evaluated by OMB staff, and a final budget request is prepared. Agency budget requests may be modified by OMB, in accordance with Administration priorities and overall budget limitations. The final budget is called the President's Budget Request (PBR) and is submitted for consideration to Congress, generally in the first week of February, although the actual submission date can vary. Submission of the PBR to Congress commences the entire process for the following year, and OMB continues to work with the various offices and agencies to plan and prepare the following year's budget request. Once Congress receives the President's Budget Request, it is divided up by agency and distributed to the various committees in the House of Representatives and the Senate with responsibility over the various agencies and offices. These committees will evaluate the budget requests and generally make modifications. The various House and Senate committees also conduct their own separate and parallel budget process. At the completion of the budget process by the House and the Senate, separate budgets that do not agree generally result. A joint committee made up of members from both the House of Representatives and the Senate, called the Conference Committee, is established and will meet to reconcile the differences in the two budgets. Once they agree on a budget compromise, both chambers must vote to approve the new budget. The process will repeat until a final budget is approved. This budget will then be submitted to the President for his approval and signature. The President has the option to either accept or reject the budget he receives from Congress. If the President does not agree with the new budget, which may have little resemblance to the budget request he submitted, he can veto it, which will result in the budget being returned to Congress for reconsideration and modification. Under these circumstances the entire process is repeated by Congress and a new and/or modified budget is submitted to the President. The budget, once approved and signed by the President, becomes public law and provides the funds for the operation of the government for the fiscal year.

Ideally, the budget will become law before the start of the fiscal year. However, it often happens that agreement on the budget does not occur until well into the fiscal year. When this occurs, the government has no funds on which to operate and may be closed down for normal business. Congress, in this case, will often pass a bill called a Continuing Resolution (CR), which permits the government agencies to operate, based upon the previous year's budget. Once the President signs the CR bill, it becomes public law, and the government is authorized to expend funds to continue to operate. Generally, a CR is passed for only a short time period, which may be a week or two, and for the purpose of permitting Congress and the Administration time to work to negotiate and compromise, and pass the fiscal year budget. Further extensions are possible with the passage of additional CRs.

6.6.2 University Policy Regarding Research Grants and Continuing Resolutions

Although the fiscal year begins on October 1, it sometimes happens that the fiscal year budget is not finalized into law until the end of the calendar year, and sometimes into the next year. When these delays occur, agency research budgets are placed under significant stress. Grant funding increments, in particular, are often delayed and researchers are faced with a disruption in their research programs. Delays in receiving expected funding increments are treated in various ways by universities. Some universities will permit research work to continue and will permit funding expenditures as if there were no funding increment delay, subject to overall funding limitations. That is, they will permit expenditures as long as the overall approved budget limits are not exceeded. In effect, the university permits the research to continue without disruption, supported with university funds, which is managed as deficit funding on the grant. Once the expected increment is received, the negative balance on the grant budget will be cancelled by the newly received grant funds. This approach has some risk, since the university has no guarantee that the expected incremental funds will actually be received. The Terms and Conditions that are included with the grant always include the statement "... subject to the availability of funds." This is a downside of the incremental funding procedure, and the statement means that the government agrees to provide the requested funds, if they have funds available in their current fiscal year budget. If the funds have not been provided to the agency, they are not obligated to provide the increment to the university. Although this situation rarely, in fact, occurs, it is possible, and the funding agency will have the option to cancel the grant. For research that is satisfactorily progressing, cancellation of the grant is not common, and program managers tend to fund existing grants and provide the budgeted increments for existing grants before making commitments for new grant initiations. Nonetheless, the potential cancellation of a grant for budget reasons poses a threat and, for this reason, some universities will not permit expenditures of grant funds if an increment has not been received. When this occurs, faculty researchers are placed in a difficult situation since they will need to devise methods to maintain their research and meet obligations until the new grant increment is received.

6.7 Research Funds Management by Program Managers and Program Directors

Once a program manager has decided to support your proposal with a research grant, they will generally determine how much funding is required to perform the research from the initiation date of the grant and through the end of the current fiscal year. This amount will generally only be a fraction of the total first-year funding requested since the first-year performance period generally does not coincide with the fiscal year budget period. The program manager may contact you and request that the budget sheet in your proposal be modified to satisfy the fiscal year funding issue. Since the general performance period for a research grant is normally three years, the actual performance period most likely will extend over four years due to the partial year funding, and the funds not provided in the first year, due to the grant initiation date, will be provided in the fourth year. In this way, the entire three-year performance period will be funded, but it will be spread over four calendar years in order to match the three-year funding to the fiscal year. An advantage of the incremental funding process for the program manager is that it permits them to fund more proposals than would be allowed if they provided the full proposal requests immediately. However, since the next increment is due the following year, the amount of funding for the initiation of new grants the following year is reduced. In fact, the majority of a program manager's core funds are generally committed to research grants that have been initiated in previous years, and most program managers will have a limited amount of funding to provide for new grants. This is one of the main reasons that the competition for the initiation of new grants is so intense. The problem becomes even more difficult in years when agency budgets are constant from the previous year, or even worse, reduced from the previous year's amount of funding. For the latter situation, very little to no funds may be available for new grant initiation. In fact, the program manager may need to let an existing grant terminate in order to have funds to initiate a new grant. This is one of the major reasons the academic community routinely supports increases in US government research agency budgets. In fact, in times of increasing research budgets, the academic community profits in a significant way since there is enhanced opportunity for new grant initiation.

The possibility of a follow-up grant from NSF is more complex than for the mission agencies. First, the NSF does not provide for follow-up grants, and continuation of the research supported on one grant with award of a follow-up grant for continuation of the same work is not possible. The only way to secure a follow-up grant is to submit a new proposal for a new project that is based upon research performed under the previous grant. In the new proposal the previous work will be described and the results obtained presented. However, the new proposal must focus upon the new work that is being proposed, and the new work must be clearly distinguished from the previous research effort. The new proposal will be submitted and will go through exactly the same review and evaluation process as all proposals submitted to the NSF. In this regard, the new proposal is in competition with all proposals submitted. The only advantage associated with the new grant proposal is that the previous work that was performed and reported permits a stronger proposal to be submitted since the new work is supported by the results of the previous work. Reviewers will consider the results of the previous work to be evidence of the likely potential success for the new proposed research, and are likely to consider this a positive factor in their evaluation. In fact, many researchers successfully receive subsequent grant funding, sometimes extending over many years.

The funding mechanism at the NSF is slightly different from that employed by the mission agencies. In the standard grant procedure at the NSF (called a Standard Grant), a three-year budget is normally approved. The entire three-year grant budget will be applied to the current-year budget allocated to the particular division providing the funds. The funds will be placed in an account that your university can use to support your research grant budget requirements. Since the entire three-year budget is available, there is essentially no risk to your grant funding for the duration of the grant performance period, assuming the work is proceeding satisfactorily. Delays due to Congress and the President arriving at an agreement on the Federal budget, which may include the passage of a Continuing Resolution, etc., will not cause any disruptions in your research funding since the grant funds have been secured for the duration of the grant performance period from the fiscal year budget in which they were committed. United States government agencies always have the option to cancel research grants, but only for specific purposes related to non-performance issues, and other extreme circumstances.

The NSF does fund a certain fraction of their awarded grants on an incremental basis, similar to mission agencies, using a process called a Continuing Grant (CG). This process functions in essentially the same manner described for the mission agencies. However, most divisions keep the number of their CGs to a minimum and attempt to minimize the total monetary value of the CGs, which is generally a relatively small fraction of their overall budget. The CG mechanism is useful to the divisions for management of their budget, particularly at the end of the fiscal year, when it is necessary for them to zero the budget. That is, all agency research funds are required to be committed by the end of the fiscal year, and the CG process is a valuable tool to accomplish this, since the amount of funding provided for grants funded by this mechanism is variable. The CG process also permits an increase in the number of grants that can be funded. An advantage of the NSF providing the majority of their grants by means of the standard grant procedure is that each fiscal year they will have most of their allocated budget available to initiate new grants. This is one of the reasons that the NSF has become a very popular funding source, particularly for new faculty researchers.

6.8 Professional Networking

Professional networking is an extremely important factor in building an academic research career. Actually, this is an understatement since networking is important for success in essentially any career, not only an academic research career, and particularly, in any scientific or technical profession. What does "networking" mean? Networking is actually nothing more than interacting and communicating with your professional colleagues on a regular or periodic basis. In this process you'll make many friendships that will last for long periods of time, some for your entire professional life. You've already started this process by graduating from your university with an advanced degree. Some of your fellow classmates and colleagues, particularly those with whom you developed friendships, can be a good start on building your network. You should make an attempt to stay in touch with these people, particularly those with technical interests similar to your own. As time progresses, you are likely to meet these people at technical conferences and meetings, and they can be a good resource to meet other people. Also, over time you're likely to meet the same people in many different forums as both their and your careers advance. You'll find the positions, locations, and titles change, but the people remain the same.

Building a professional network requires making contact with many people. The best opportunity to do this will stem from your place of employment and related organizations, and your professional activities. These two activities are related, but distinctly different. In your employment you will interact with many people on various levels of the professional organization, and you will have many colleagues with whom you will interact on a daily or periodic basis. As time progresses, you and your colleagues are likely to move from one company or organization to another, and sometimes relocation to organizations across the country, or even relocation to other countries, may result. However, these people, particularly those with whom you have similar technical interests, are likely to attend technical and professional conferences and meetings in your specialty interest areas and you will have the opportunity to maintain contact with them. You are very likely to meet these people at these conferences and meetings and have a chance to stay in touch and continue your relationship. These networking opportunities, which should be pursued and maintained, can have many benefits, extending from personal to professional activities. These relationships can even result in future employment opportunities. Also, continuing your relationships will likely result in significant growth in your network as you meet and develop relationships with people you meet through your colleagues.

Professional organizations, particularly through their conference and meeting activities, also offer a very significant opportunity for building your network. Virtually all professions are supported by professional societies made up of people who work in the profession. For example, people working in electrical engineering areas are supported by the IEEE, professionals working in materials science and engineering are supported by the Materials Research Society (MRS), physicists are supported by the Applied Physics Society (APS), and chemists are supported by the American Chemical Society (ACS). These professional societies and organizations sponsor numerous conferences and meetings throughout the year. Some of these conferences are large and include many technical areas, and some of the conferences and meetings are relatively small and dedicated to a single subject. Attendance at these meetings can be large, with hundreds or thousands of participants for the general conferences, or limited to a relatively few participants for the specialty topic meetings. You will need to determine what conferences and meetings are most appropriate for you, based upon your technical interests, and then make every attempt to attend and participate in these meetings. Attendance at these meetings presents an extremely fertile opportunity for you to meet new people working on topics of mutual interest and with whom you can discuss problems and research approaches. Many of these people will become your colleagues and some will become collaborators.

You'll also find that many funding agency program managers and program directors are likely to attend these meetings. This offers an excellent opportunity for you to meet and build relationships with them. Meeting and talking to them during coffee breaks, etc., presents a good opportunity for initial contact. However, you'll probably find that many of these people also serve on program committees and meeting organization and planning committees. Your participation in these activities presents an excellent opportunity to get to know these people on a deeper level, and to further develop relationships. You should volunteer your time and effort and get involved in as many of these activities as your schedule permits as these activities present an excellent opportunity to continue to build your network of professional colleagues.

6.9 What We've Learned

In this chapter we've discussed the very important issue of how one goes about identifying and making contact with the program manager or program director most appropriate for his or her research area. We explained the qualifications of the program managers and program directors, and how they are selected to serve in the respective positions. The role of the program manager or program director was described, including their involvement in program development, and how they are evaluated by their organizations. We also described several US government grant funding agencies, and described the various methods they have to advertise their research opportunities. We then presented various techniques that could be used to communicate with appropriate program managers and program directors within the funding agencies. It is emphasized that personal contact and communication with the program managers or program directors is fundamentally important in order to build their confidence in you and your research. We also discussed the concepts of white papers and year-end money, followed by some considerations regarding how the agencies evaluate proposals. We also addressed both the funding agency, and university research grant, management processes. The chapter concluded with a discussion of the need to build professional networks, and mechanisms that can be successfully employed for this purpose.

7 The Proposal

Preparing and submitting a research proposal is, simply stated, the most important aspect of building an externally funded academic research program. It's one of the most important activities associated with holding an academic position in a research-oriented college or university, and achieving success in an academic career is very much tied to the proposal preparation and submission process. This single activity will dominate your academic career, and for as long as you remain active in your profession. That is, your need to organize, prepare, write, and submit research proposals will never end. You will be involved in this process until you leave your academic position, either through your retirement, or to take another job. For this reason, you need to learn the basics of preparing a quality proposal that will give your ideas the best chance for obtaining financial support. You will learn that proposal writing is a skill that can be learned, and that, as you gain experience, the time and effort required to prepare a quality proposal will decline.

In this chapter we'll discuss the proposal and examine not only what it should contain, but also how to present the material in an effective style. Your ideas need to be clearly stated, but in a brief and informative manner. You don't want reviewers to have to struggle to find out what you are trying to present, or to learn what is new or novel about your ideas or your approach to the research problem you are addressing. You want the main ideas to be readily apparent, and to basically "jump" off the page to the reviewer. You want to make the reviewer's job as easy as possible. For this reason, many experienced proposal writers will use techniques to make critical statements impossible to miss. For example, many authors will place short, but critical, statements in bold type. We'll discuss these, and other, techniques as we go through the basic structure and presentation of the proposal.

The proposal is the main official mechanism for you to directly, and legally, compete for research funds, no matter the source of the funding. Proposals submitted to US government and private funding agencies will have essentially the same basic content, although the proposal format and certain items and required declarations and statements may differ slightly. A research proposal is, by definition, a legal document in which you make an offer to perform a defined research effort in exchange for financial support. The agreement is actually made, from a legal perspective, between your home institution and the funding agency, and your home institution research administration will provide the legal signature required for the proposal submission to be officially received and entered into the evaluation procedure. The research office, generally called the Office of Sponsored Research, or some other equivalent title, will actually submit the proposal to the sponsoring organization, and once you submit the proposal to your institution's research office, you are not directly involved in the process, although you will be informed of the proposal status. Since the proposal must be reviewed and approved by your home institution, you need to be aware of deadlines, and make sure you submit your proposal through your home institution's approval process in a timely manner to ensure that they have adequate opportunity to review your proposal and perform the approval process and submit the proposal in time for it to be received by the funding agency, consistent with their submission deadlines. The necessary lead time for your home research administration's review and authorization will vary from institution to institution, but will generally be on the order of two to five days, although more time may be required during periods when they are experiencing heavy proposal submission volume. This often occurs as the due date for popular proposal submission windows approach. The guidelines for the exact time for submission will be posted on your home institution's research administration website, and the information is readily available from them. Your research administration's review is not generally a technical review, but a review to check your budget, including items such as correct labor categories and rates, cost-sharing commitments, permitted travel obligations, indirect charges, and to make sure your budget is consistent with the agency guidelines. They will also check to make sure any required declarations and statements are included.

Submission of the proposal to the funding agency will follow formal procedures, which are defined by agreement between the funding agency representative and your institution's research office. The requirements of the funding agency will be indicated and listed in the Call for Proposals, or other research opportunity announcement, and all proposals that are submitted by all proposers must adhere to the stated format and include requested items, declarations, and statements. Failure to follow the stated format, or to neglect to include necessary declarations or statements, will result in the return of your proposal to you without review. That is, once the proposal is received by the funding agency by the stated due date, it will be quickly scanned for format and content, and if any items are deficient, the proposal will not be entered into the evaluation process or read. Therefore, you need to carefully read the proposal solicitation and make sure you follow all guidelines. Submission of a research proposal to a funding source is, as stated, a legal procedure that is governed by applicable state and federal regulations and agreements.

Research funding sources will announce their funding opportunity, and you will prepare and submit a proposal to them, following criteria described in their Call for Proposals or other research funding opportunity announcement. The funding opportunity, which is open to all qualified bidders (proposers), may be for a specific research topic with a defined submission date, or it may be recurring and open for proposal submissions at any time during the open period. Whatever the specifics of the funding opportunity, you will devote significant time and effort on proposal development. In fact, you'll find that you will, most likely, and particularly in your first few years, spend almost as much time on proposal development as you do on any other single activity. Your home institution, of course, recognizes the importance of this activity, and many colleges and universities offer mentoring in proposal preparation in order to help new faculty members learn and improve their proposal preparation skills, and to facilitate the process. You need to become an expert proposal writer as the difference between submitting a successful proposal and being declined will depend not only upon the quality and contents of the research objectives and plans presented in the proposal, but also how well the proposal is prepared and written. It cannot be overemphasized that writing skills and proposal organization are extremely important!

You need to learn what to include in a proposal and how to effectively present the information in order to receive favorable reviews. There is much skill and art in developing a successful proposal. However, there are several fundamental principles associated with preparing a proposal that will enhance your success. We will discuss these principles in this chapter, as well as the content and elements of an effective and well-structured proposal. You will find that, as you advance in your career, and go through the proposal preparation and submission process, your skills will improve and that the time and effort required for you to prepare a strong and competitive proposal will decrease.

7.1 Some General Comments From an Experienced Proposal Reviewer

Before we investigate the proposal preparation process in more detail, I can offer some general comments from my personal experience while serving in numerous and diverse proposal evaluation capacities. As a mature proposal reviewer with over four decades of experience reviewing proposals for a wide range of government and private research sponsors, I can tell you that many proposals are very poorly written, even by mature and successful researchers, even though they may contain interesting information concerning a specific topic and may include novel approaches to complex problems. However, if these ideas, concepts, and approaches are not clearly defined and explained, the proposal reader will not gain a clear understanding of the proposed research. Although a poorly written proposal may not necessarily limit a well-known and established researcher from gaining funding, the same cannot be guaranteed for a new or relatively unknown researcher. In these cases, the proposal writers do themselves a disservice, and place the potential of funding success for their research at a significant disadvantage by the failure to effectively communicate their research ideas and approaches. Poorly written proposals tend to receive low ratings, even if the ideas contained in the proposal have merit and warrant being funded. Conversely, well-written proposals that very clearly and effectively present the research problem and include a solid scientific approach to address the problem often receive high ratings, even if the significance of the research problem is not well established, or the problem is not considered of the highest priority to the research community or the funding agency.

The principles for a good proposal that I list and describe in this chapter are derived from many years of service as a US government program manager, as well as service as a proposal reviewer for a wide variety of US government, foreign government, state government, private industry, foundations, and other research-funding agencies. I have also served as Chair of numerous proposal review panels for various government and private funding agencies and organizations, and have worked in management positions for US government funding agencies where I directed program managers and had management responsibility for their proposal review and evaluation activities. Along with my own experience both as a reviewer and also as a program manager directly involved with the proposal preparation and evaluation process, I have also observed how my proposal reviewer colleagues, from both government agencies and private institutions, perform their reviews and what sort of comments they tend to offer on proposals while serving on proposal review panels. I have personally reviewed a large number, probably approaching or exceeding 1000 research proposals, and sometimes with little time allocated to the review. For example, while serving as a program manager, I once was required to review and evaluate about 60–70 SBIR Phase I proposals for a US Department of Defense Agency over a time period of three to four days that extended over a weekend. I took the proposals home with me and spent the two-day weekend on my living room couch doing essentially nothing else but reading and evaluating proposals. Needless to say, under these conditions, only a limited amount of time can be devoted to each proposal in order to meet the schedule. Brief, well-written proposals have a much greater chance for receiving good evaluations, while lengthy and poorly written proposals stand a high probability of receiving low evaluations. This particular review activity is probably a little extreme, but it demonstrates the pressure that can be placed upon reviewers. However, the need to evaluate a significant number of proposals in a limited time period is common for funding agency program managers and proposal reviewers. As a researcher, you want to make the job of the reviewer as easy as possible, and this can be best accomplished by learning to present your ideas in a clear and concise manner.

7.2 Who are Proposal Reviewers and How Does the Proposal Review Process Function?

Since the proposal review and evaluation process is so critical, you're probably wondering who will actually perform the function. You also may be wondering if you will have any ability to select or suggest people that you would like to serve as reviewers on your proposal. The answers to these questions, particularly the second, can be complicated, but in general, the answer to the second question is "no," you will not be able to select the reviewers for your proposal. Also, you will not know the identity of the reviewers. The NSF does permit you to identify people that you feel are qualified to review your proposal, as well as provide names of people that you specifically do not wish to serve as reviewers for your proposal. However, in my experience, the list of potential reviewers is seldom used, unless they are already in the qualified reviewer database. The list of people you are not comfortable with to serve as a reviewer for your proposal will not be selected. Mission agencies such as the DOD, DOE, NASA, etc., generally do not ask for a list of potential reviewers since they already have their own mechanisms and procedures for identifying qualified reviewers.

Essentially all grant funding agencies make use of a peer-review process where the identity of the proposal reviewers is blind to the author of the proposal. However, you can be assured that they will all be competent and technically proficient and qualified to perform the review. So, who are the reviewers and how are they selected? Proposal reviewers are your professional colleagues, and they are derived from a variety of organizations, which could be other academic institutions, government agencies and laboratories, industrial organizations, and research institutes. Although you will not know their identity, their evaluation of your proposal and their comments will generally be provided to you. Also, you may request a debriefing on your proposal from the program manager or program director after the review and evaluation process is completed, and the program manager or program director will generally provide you with additional information regarding the review, sometimes indicating issues that were brought up during the panel deliberations.

The proposal reviewers will be selected by the program manager or program director based upon their expertise in the subject of the proposals that they are asked to review. Although all of the reviewers will be familiar with the technical area of the subject of your research proposal, some of them may lack in-depth knowledge of your specific proposal subject, and they may have only a limited knowledge of the scientific or technical details of your particular research topic or approach. Conversely, some of the reviewers will be experts in your subject and will have a deep understanding of your research topic. The program manager will generally recognize the range of expertise of the reviewers and will honor the limitations associated with certain reviewers. They will weigh the comments from the various reviewers in a balanced and appropriate manner.

If a panel is evaluating your proposal, the program manager or program director will generally assign your proposal to the reviewers with the most expertise in your specific topic, and will ask one of the most knowledgeable reviewers to lead the evaluation procedure. However, since you can expect that at least some of the reviewers that will be assigned to evaluate your proposal may have only limited knowledge of your subject, you need to learn to write your proposal for readers who are technically proficient, but not necessarily knowledgeable of the intimate details of your topic. You also need to include sufficient in-depth details for readers that have significant expertise in your subject. There is an art to writing a proposal with enough background information for general readers, while also including sufficient in-depth material to demonstrate what is new or novel about your ideas and approach to the research problem you are addressing. The latter is very important since you are attempting to convince them that your ideas are worth the financial investment that you are requesting. Also, it is now common for funding agencies to limit the number of pages of technical information that can be included in a proposal, and this page restriction mandates that an appropriate balance of background and new material be presented in an effective manner. This is, in fact, one of the most difficult aspects of writing and preparing an effective and successful proposal. You will find that your writing skills improve as the number of proposals you prepare increases.

7.2.1 The NSF Proposal Review Panel Procedure

Before discussing the proposal review process, it's helpful to gain an understanding of the people that will organize the review of your proposal. At the NSF these people are called program directors. These program directors consist of a mix of permanent US government employees and temporary employees serving on leave from their home institutions. The temporary employees are generally academic faculty members serving in the program director role at the NSF under the US government Intergovernmental Personnel Act (IPA), which permits the government to hire personnel from non-government institutions and organizations on a time-limited and temporary basis. The IPA employees can serve as a program director at NSF for a limited period of time, for one year at a time, and extending up to four continuous years. While serving at the NSF under an IPA agreement they remain employees of their home institutions and, in fact, continue to receive their salary and benefits from their home institution. The NSF provides the institutions with the IPA employee's salary and benefits, and at the same rate as their normal salary, etc. There is no increase or decrease in the salary, and the NSF will modify the agreement to account for salary increases, etc. Some IPA program directors serve for an additional term, and the IPA program as used by the NSF, permits an IPA employee to serve at the NSF for up to six years of the previous 10-year period. For example, a faculty member serving as a program director under an IPA agreement could serve a three-year period as program director, and then return for an additional three-year period as a division director. Other possibilities also exist.

The NSF primarily uses a panel review process to evaluate proposals for both their open window opportunities, as well as for directed solicitations offered under a Dear Colleague Letter. Once the proposals are submitted, generally through the NSF FastLane or Grants.gov websites, the proposals are scanned by the system to make sure they adhere to the proposal submission guidelines. That is, the system will automatically check to make sure certain information is contained, and that the proposal length does not exceed the maximum permitted number of pages. If the proposal does not meet the guidelines, it will be returned to the submitter without review and will not be further processed. The proposals that meet the guidelines are divided up by subject area indicated in the proposal, and then distributed to the program directors. The program directors will scan the proposals for content and to make sure they address the scientific, technical, or educational research area appropriate for the program director's area of responsibility. If the program director feels that a certain proposal is not appropriate for his or her panel, he or she may negotiate with another program director in another area, and the proposal may be transferred to a different area, or possibly, in rare circumstances, a different division or directorate.

Each proposal review panel will consist of a number of members, ranging from 12 to 15, although the number will vary from panel to panel, depending upon the subject area. However, panels that exceed 20 members are rare and discouraged, as they become too large to effectively manage. The program directors have the responsibility to recruit the scientific and technical experts to serve on the proposal review panels and the choice of who will participate on the panel is the program director's decision. In order to assist and facilitate this process, the NSF maintains a database of qualified reviewers, listed according to professional area and topic. The program directors make extensive use of this database to identify reviewers appropriate for their specific panel.

They may also recruit people who they know either personally or professionally to be experts in the scientific or technical subject of the proposal review panel.

Once the members of the review panel have been identified and successfully recruited, the program director will assign each proposal to a number of reviewers, generally ranging from four to six, and sometimes more. They will be contacted by email, and provided with information regarding the panel area, panel ID, and password information to log onto the NSF FastLane system. Once in FastLane, they have access to information regarding the panel, including travel and logistics details, proposal assignments, and the actual proposals that will be reviewed by the panel. Their specific assignments will also be indicated, and the reviewers will download the proposals, read and evaluate them, and enter both their summary evaluation and specific comments into the FastLane website. The summary evaluation consists of a grade, such as E for excellent, VG for very good, G for good, F for fair, and P for poor. All proposal reviews are to be completed before the time the panel meets, generally at or near the NSF office in northern Virginia, near Washington, DC. During the panel review, the grades and comments entered into the system during the preliminary evaluation process serve as the starting point for the discussions, but the panel will determine the final evaluations, and the initial grades may change during the panel procedure when the final proposal evaluation and grade are determined.

Program directors at the NSF do not generally personally evaluate proposals, and most do not actually personally read proposals, with the exception of the executive summary, along with a possible scan of the proposal to ensure they understand the topic. This is important since they are responsible for assigning the experts who will perform the detailed review, and they need to make sure they assign the proposal to reviewers with expertise in the subject. The primary function of the program director is to organize and manage the proposal review, to make sure the proposals are fairly evaluated by reviewers with expertise in the subject area of the proposal, and ensure a fair evaluation process. The program directors lead the panel in determining the final rank ordering of the proposals in any given competition.

The vast majority of proposals submitted to the NSF are evaluated by proposal review panels, where the reviewers will evaluate the proposals, discuss each proposal, and rank them, generally into one of three categories, consisting of: (1) Fund; (2) Fund, if possible; and (3) Do not fund. Often, after further discussions, this list will be reduced to two categories of (1) Fund and (2) Do not fund. The program director will have overall management responsibility for the panel, but will delegate a participant of the panel who is expert in the research area of the proposal to lead the discussion and evaluation of each proposal. Not all review panel members will read all proposals being evaluated by the panel. Each review panel will normally consist of about 12 to 15 reviewers, and each will possess either specific and detailed knowledge of the panel subject or, at least, a general knowledge of the research area. Each proposal will be assigned to a subset of the panel, generally consisting of about four to six members, and sometimes more. One member will be identified as the lead reviewer, and one member will be listed as the scribe. The lead reviewer will assume responsibility for leading the panel discussions on that particular proposal, and the scribe will record significant comments or issues that arise during the panel discussions. The scribe may or may not personally read and evaluate the proposal. However, each proposal is guaranteed to receive at least three in-depth technical reviews, and most will receive four or five. Each member of the panel will serve as lead, scribe, and reviewer on several proposals, with the exact number dependent upon how many proposals are to be evaluated in the panel. The program director will attempt to distribute the duties equally so that all panel members have a balanced work load.

During the review and evaluation procedure, the lead reviewer will normally manage the discussions of the proposal, and each member assigned to review the proposal will contribute their comments, generally starting with the comments they entered into the system in the review they performed before the panel meeting. They will also indicate their overall grade for the proposal. Each proposal is assured of receiving detailed reviews from the panel participants assigned to the proposal. In the proposal discussions, the members of the panel will determine a single final grade for the proposal, and provide comments concerning the proposed research. If the grade determined for the proposal by the entire panel during the discussion doesn't agree with the individual reviewers' grades, they will often change their grades to be consistent with the overall panel grade. However, the individual reviewer grades will still generally vary and will not necessarily be the same as the overall panel grade. The scribe will produce a written summary of the proposal review, based upon the comments made during the evaluation procedure. The final summary is entered into the FastLane system, and is approved by all members of the panel, in order to assure that the scribe has accurately captured the panel evaluation. Panel members not specifically assigned the proposal may also offer their comments, and many do so if the proposal addresses a subject with which they are familiar. However, their comments may not be directly recorded or entered into the written summary, unless they enter themselves into the system as a reviewer, which they are permitted to do. At the end of the panel all panel members must concur with the final evaluations and grades entered into the FastLane system, and the system will not permit the panel review to be concluded until all members of the panel have done so.

During the panel review and evaluation procedure, the program director will normally group the proposals into those that have received all excellent or very good grades in the pre-panel reviews, those that have received all fair or poor grades, and the proposals that have received mixed grades. Those that have received all fair or poor grades may not receive detailed discussion, and may be quickly moved into the "Do not fund" category. Likewise, those that have received all excellent or very good grades may receive only brief initial discussion, and will be set aside for further discussion after the other proposals have been discussed. Most of the panel discussion time will be devoted to proposals where there is not a strong consensus concerning into which category the proposal should be placed. The program director will ensure that each proposal receives a fair and equitable evaluation, and there may be more than one round of discussions regarding certain proposals, particularly if controversy occurs. The final product of the panel will be a list of a small number (generally no more than one to four) of proposals that the panel recommends for funding, and a list of the remaining proposals that the panel recommends to be declined. After the panel concludes its duties and departs, the program director will make the final determination of the list of proposals that he or she will recommend to be funded, primarily based upon the final panel rankings of the proposals, and will submit the recommendation to the NSF division director for approval. The program director will generally select a particular proposal to be funded from the list of proposals that have received the highest ratings. However, they may not necessarily select the proposal that received the highest rating due to considerations of subject area, emerging technologies, and other factors that the program director feels warrant consideration. The division director will give the final approval for funding, based upon their review of the ranked proposals and the division budget. The division director has access to all information and reviews entered into the FastLane system and has responsibility for concurring with the program director's recommendations.

Some highly ranked proposals may not be selected for funding, due to lack of sufficient available funds. These proposals would, of course, be declined. However, it is also possible that a highly ranked proposal not initially selected for funding could be held by the program director for a period of time, with the potential that the proposal could be selected for funding if additional funds become available later in the fiscal year. In these cases, the proposal writer will generally not be informed of any decision regarding the acceptance or declination of their proposal until a decision is actually made. This process and lack of communication can make the proposal writer very nervous regarding the status of their proposal. However, it sometimes is true that no news is good news, and patience on the part of the proposal writer is warranted. The results of the panel review are generally quickly disseminated, and the principal investigators and their institutions are promptly notified soon after the final decisions have been made. If you do not receive notification of the final status of your proposal in a timely manner, your proposal may be in the hold category. This situation is, in fact, common since additional funds generally become available at the end of each fiscal year, a situation that occurs since funds appropriated by Congress need to be entirely spent and the budget zeroed at the end of each fiscal year. If funds are not spent by the end of the fiscal year by a US government agency, they are returned to the Department of Treasury and their budget could potentially be reduced by that amount in the following year. This, of course, provides high motivation to US government agencies to completely spend their budgeted funds each fiscal year.

At the NSF, if the appropriated and allocated budget to each directorate and division has not been entirely spent during the fiscal year, some funds may remain towards the end of the fiscal year, and this generates the opportunity for some additional highly rated proposals to be selected for funding. The NSF manages this process on the directorate and division level, and at the end of the fiscal year, the directorates will normally perform what they call a "sweep" of unspent funds in each division's budget. These are funds that have not yet been committed to fund proposals. Once the sweep is concluded, the directorates will often send out a message to each division director announcing that additional funds are available for proposal support. This provides the opportunity for additional highly rated proposals to be funded. Many program directors who have held one or two proposals that they feel should be funded will submit these proposals to the division director for consideration of funding. The division director will generally make use of this procedure to fund areas that he or she feels important and deserving of additional support.

Proposal evaluation panels are fundamental to the review and evaluation procedure, and you should become involved in the proposal review process as soon as possible. As your career progresses and you publish and present your work in professional forums, at some point you will likely be asked to serve as a research proposal evaluator and reviewer, either for a single proposal, or on a research proposal evaluation panel. The NSF, in particular, is always looking to expand their base of proposal reviewers. You should accept as many requests as possible to serve on proposal evaluation panels. The experience is very interesting and enjoyable, and is invaluable for learning the content and structure of both well-written and poorly written proposals. You will also learn how research proposals are evaluated, and this will help you to develop your proposal writing skills. Also, service on a research proposal evaluation panel offers an excellent opportunity to make personal contacts with both NSF program directors and your professional peers and colleagues. In order to volunteer as a proposal reviewer, you can contact an NSF Program Director and request to be added to their list of reviewers. They will likely respond favorably and ask that you complete a form that asks for your contact information, area of expertise, and your credentials. Once they receive this information, they'll consider your request, and then add you to the list of reviewers in the database. Once you serve on a panel, you'll likely be requested to serve on additional panels in the future.

7.2.2 Mission Agency Proposal Review Process and Panels

Program managers at other government agencies, such as the DOD, DOE, NASA, etc., consist primarily of permanent US government employees. The mission agencies also make use of IPA employees obtained from non-government organizations, such as academic institutions and industrial organizations, but in much smaller numbers than used at the NSF, and only for specific reasons. That is, they do not generally recruit outside people to serve in their institutions as program managers under IPA agreements unless there is a specific need or reason to do so. The program managers in certain mission agencies may be scientists or engineers on detail from other US government organizations. For example, many of the program managers serving at DARPA are employees of other DOD organizations, such as ONR, AFOSR, ARO, etc., and will return to their home organization once their detail period has expired. A primary purpose of the detail is to acquaint the program manager with DARPA programs, and attempt to bring enhanced alignment between the DARPA programs and their home organization plans and programs.

The program managers in the mission agencies operate in a similar, but slightly different, manner from program directors at the NSF. For example, the majority of program managers at most government agencies, other than the NSF, are permanent, professional government employees and, as discussed in Chapter 6, are building and financially supporting a research program to achieve an overall goal, consistent with their office and agency mission. How they handle your proposal will vary, depending upon how it was received. If the proposal was submitted in response to a specific Call for Proposals on a defined subject, a panel of experts, generally selected by the program manager, will most likely evaluate the proposal. In this case, the panel evaluation will proceed in a manner analogous to that discussed above for NSF panels. However, the program manager may, or may not, actually read your proposal personally, depending upon their level of interest in the subject of your research. You want them to personally read your proposal so you need to communicate with them before submitting your proposal to determine their specific interest areas. The program manager will generally actively participate in the panel review and will often make specific comments regarding proposed approaches, recent developments, and other factors. Keep in mind that they are focused upon accomplishing an end goal, and they are looking for scientific and engineering contributions that will assist them in their overall objectives. The final selection of the proposals that are to be funded will be made by the program manager, and possibly their office or agency colleagues, based not only upon the review panel rankings, but also upon their overall and specific program goals. In some cases, they may select a lower-rated proposal to receive funding over a higher-rated proposal, where they based the final decision upon the technical subject, their desire to have research performed on a specific subject, and on how well they feel the proposed research coordinates with other research that will be performed on the project or development area.

7.3 How Experienced Reviewers Read Proposals

Professional program managers and program directors receive and read a large number of proposals. They also recruit a large number of experienced reviewers to assist them in evaluating the proposals they receive. Experienced proposal reviewers become very skilled at reading a proposal and can quickly and effectively determine what is being proposed, and the significance of the approach and research topic being proposed. Most experienced reviewers will initially speed-read a proposal, skimming over certain information, while seeking the real "meat" of the proposal and attempting to identify exactly what topic is being addressed and what advance is being proposed. For this reason, you want to make sure your specific contribution and approach are very clearly and precisely stated. Once they have a good idea of what you are proposing they will also look for information relative to: (1) the fundamental research problem, including previous work done on the subject and the current state-of-the-art; (2) the approach to be pursued, including what is new or novel about the proposed approach; (3) supporting data and previous results; (4) the proposer's access to equipment, instrumentation, laboratory facilities, and other resources necessary to perform the research; and (5) the credentials of the principal investigator (PI). If any of these items are not included or adequately addressed, the omission will be considered a deficiency and the proposal rating will be downgraded. The proposed budget will also be examined, but generally only in a brief manner. Generally, reviewers will look to see if overall budget requirements are met, if the PI has included adequate funds for their time, student support, and if they have requested reasonable amounts for supplies, travel, etc. However, the budget details are generally left for the agency program manager to evaluate, and the proposal reviewers will make comments only if something looks inconsistent with overall program goals, or something that appears inappropriate is included. Experienced proposal reviewers will scan information generally considered as "boiler plate" information. That is, standard information regarding laboratory and computer facilities, the university support for research, and other generic information concerning the university or its desire to build their program. While this information may include some valuable factors, most program managers and reviewers will only scan the material, unless something catches their eye that they feel affects the proposed research or the ability of the PI to perform the research.

The reviewer will generally form an initial opinion and draft rating for the proposal from this initial reading and evaluation, and the other information included in the review will be left to a more-detailed final review after the initial opinion has been determined. The initial reading will direct the reviewer towards subjects and topics that require a more detailed reading in order to fully understand the proposed research. Information related to items identified in the initial scan or items that piqued the reviewer's interest will be the subject of the more detailed evaluation, particularly if the reviewer felt he or she was confused over a subject and thought that a critical detail was missed or not understood. The detailed review will focus in these areas. Once the reviewer feels that he or she understands the proposal, the review will be terminated, and this information will then be used to determine the final rating and summary comments. Many times the final rating is primarily based upon the initial reading, particularly for either very well-written proposals focused upon timely research topics, or for very poorly written proposals that present little new information or are not concerned with a research topic of interest. The latter issue is more common for proposals submitted to mission agencies, where the opportunity is directed to specific defined research topics, than for proposals submitted to the NSF, where proposals on a wide diversity of topics are generally considered acceptable for a given opportunity. However, even the NSF will reject proposals for lack of interest reasons when the proposal is submitted in response to a defined research topic and the proposal topic does not fit well within that topic.

For these reasons it is very important that the proposal be clearly and concisely written. Also, major items that are the focus of the proposal need to be clearly defined and simply and explicitly stated. Overall, and for a first read, the reviewer may spend only 10 or 15 minutes on your proposal, or they could spend an hour or more. Less time is required to review very well-written, and very poorly written, proposals. These two categories of proposals are generally easy to determine, and the evaluations will generally be quickest and easiest to perform. Proposals that do not clearly present the case for why the research is worth funding, or that confuse the reviewer, may take longer to review. Proposals that require that the reviewer spend an hour or more to review and evaluate often will not be highly rated, although this will vary and there are certainly good reasons a reviewer will spend extra time reviewing a proposal, particularly if they feel it may contain a novel concept or approach. In this case the reviewer may spend extra time to make sure their evaluation is, in fact, correct. However, most reviewers will not have the time to spend an hour or more evaluating a proposal when they have a significant number of proposals to review. In these cases, the proposal author, by submitting a weakly written or disjointed proposal, is gambling that the reviewer will spend the extra time to evaluate the proposal, which may require the reviewer seeking out additional information from other sources. While many reviewers will take the time and spend the effort to do this, this is not guaranteed. They may simply give the proposal a poor rating. In my experience I've seen reviewers take both options, and I've witnessed many proposals receive poor ratings, even when the proposal actually included novel results or a novel approach, but the proposal was not well written. If the reviewer doesn't completely understand the ideas in the proposal, it's very easy for them to simply conclude that "the proposal contains nothing new." I've also witnessed proposals that contained very incremental approaches and less than impressive results receive excellent ratings, simply because the proposal was very well written.

7.4 Basic Principles for Preparing a Research Proposal

As mentioned above, there is much skill and art involved in preparing and writing a high-quality research proposal. You want to enhance your chance for having your proposal favorably received and reviewed, and this can best be accomplished by learning to present your ideas in an effective, concise, and professional manner. You want to present yourself as a competent and skilled researcher with novel and new ideas that will advance your field of research. However, just being competent with novel research ideas generally isn't sufficient for success in obtaining research grants. You also need to learn to effectively present your ideas through well-structured proposal organization and presentation. We will discuss the elements of a well-constructed proposal in this chapter, but first we'll review some basic principles that will prepare you to organize and construct an effective research grant proposal.

7.4.1 Principle 1: Know Your Subject

Knowing your subject may seem to be an obvious principle. In fact, you were most likely recruited to your current academic position based upon your expertise and perceived ability to build an effective research program. You have probably devoted much time and effort to your research and feel yourself highly qualified. However, you are now competing for research grant funds on a national basis and your proposal will be evaluated in competition with other proposals submitted by both mature and new investigators. In addition to your obvious familiarity with your own work, you should be familiar with work presented in the major publications and major conferences in your field by other researchers. It's important to know the latest developments in your field and the identities of the major researchers in your field. Basically, you need to know your competition and what approaches are being pursued. Different researchers pursue different approaches, and there may be controversies regarding certain results or approaches that have been presented. If your subject contains some controversial elements or concepts, there may be panels at conferences or workshops devoted to the issue. If so, you should make every attempt to attend these meetings and participate in the discussions. You need to be aware of the scientific or technical issues and questions that have been raised, particularly if your research approach favors or follows one of the prevailing approaches. If your proposal topic is in a controversial subject, keep in mind that reviewers who favor an alternative approach may review your proposal, and this could result in your proposal receiving an unfavorable review. Therefore, you want to write your proposal in as non-controversial a manner as possible. However, demonstration of a thorough knowledge of the subject will always work to your advantage. For example, with a controversial topic, various alternative approaches or theories should be discussed in the background, along with both positive and negative factors associated with the various approaches. You should then describe your proposed research, and position your approach in a manner to help elucidate further knowledge that will help clarify the subject. In this manner you can address the scientific and engineering basis of the research without the need to take one side or the other.

7.4.2 Principle 2: Not All Proposals are the Same – Learn to Write Your Proposal for the Funding Agency and Their Reviewers

Although the vast majority of proposals that are submitted, regardless of funding agency, will have essentially the same content and the same general format, there are distinct content differences that need to be recognized. We'll discuss these content differences later. However, an often overlooked, but very important, factor to recognize is who will, most likely, review and evaluate your proposal. That is, you should anticipate if your proposal will be reviewed by professional scientists and engineers employed in the funding agency, or by outside experts. You should seek to know if there is a directed theme to the Call for Proposals, or if the funding agency is simply looking for proposals that present the best science or engineering ideas. The former is common in mission agencies, whereas the latter is common for the NSF. You should then structure your proposal and present your information with regard to the anticipated reviewers so that they will be easily able to understand your ideas. There is a difference regarding how your proposal will be reviewed, depending upon the funding agency to which it is submitted, and you should write your proposal with the anticipated reviewers in mind.

For example, your academic colleagues will review virtually all proposals submitted to the NSF, with the possibility that a small number of reviewers will come from government agencies or private companies. Since academic peers will be the most likely reviewers for your NSF proposal, you should take extra care to make sure you adequately explain the background to your research and indicate the significance of your approach and your past results, if any. You can also anticipate that your proposal, if submitted to the NSF, will be reviewed and evaluated by a panel of experts. You should assume that at least some of the reviewers will have only marginal or limited knowledge of your research topic. This situation is made quite likely by the very strict conflict-of-interest (COI) policies employed by NSF. Under these COI policies, anyone submitting a proposal to a specific Call for Proposals (called a Dear Colleague Letter by the NSF) is disqualified from serving as a proposal reviewer on any panels associated with that particular research opportunity. Since many experts will submit proposals to a given and specific research opportunity, many very knowledgeable experts are not able to serve as reviewers. You can almost be assured that the most knowledgeable experts on that particular subject will also have submitted proposals and will, therefore, not be among the reviewers. The NSF program directors will often address this situation by recruiting reviewers who are skilled in related, but not necessarily the same, areas of expertise. They will be familiar, in general, with the subject, but will not necessarily be experts familiar with details of the existing or proposed research. Also, the COI policies will not permit anyone from your home institution, anyone with whom you've served as a co-author on a publication at any time in the past four years, or anyone with whom you've served as a co-editor of a journal, compendium, or conference proceeding at any time in the past two years, to serve as a reviewer for your proposal. Therefore, it is likely that some of the panel reviewers assigned to your proposal will not have detailed knowledge of your topic, and you need to recognize this and prepare your proposal appropriately, and concisely and effectively explain your research, taking care to note why it is important, and what is new that you propose to add to the subject. You should clearly state, and indicate, new and novel results and ideas explicitly. Keep in mind that the NSF is primarily looking for the best research proposals addressing subjects in a relatively general area, and not necessarily research proposals that integrate ideas in a tightly synchronized manner from the various proposals being evaluated.

Proposals that are submitted to mission agencies, such as the DOD, DOE, NASA, etc., are processed in a slightly different manner, and scientists and engineers from agency laboratories, along with selected academic experts, will likely review the proposals. However, the academic experts will be selected based upon their specific expertise in proposal subjects, and can be expected to have detailed knowledge on your topic. Often, your proposal will be reviewed by the most knowledgeable experts in a specified subject, and particularly if those experts are also funded by the agency to which the proposal has been submitted. Mission agency program managers have a tendency to ask scientists and engineers they are funding to serve as reviewers for other proposals they have received.

Mission agencies will use both expert panels and individual expert reviews for proposal evaluation purposes. For specific topic research announcements, you can anticipate that proposal evaluation panels will be employed. However, you can expect that essentially all the reviewers assigned to evaluate your proposal will have significant or intimate knowledge of your research topic, and how your research fits into the overall area described in the announcement. However, if you submit an unsolicited proposal in response to an open BAA, you can anticipate that three to five individual experts will be asked to review your proposal. The experts will be selected from internal agency offices and laboratories, and outside academic organizations, and all reviewers will generally have intimate knowledge of your research topic. The proposals are generally sent by email to the reviewers, who will perform the evaluation at their home institutions, and then input their ratings and summary on the agency proposal website. Proposals submitted to the mission agencies should include background information, but this information generally doesn't need to be presented in the same manner as for proposals submitted to the NSF. In the background section you should indicate the general status of the research topic, taking care to note the most important developments. You want to set the scene in order to describe the advance you are proposing. You also want to indicate to the reviewers that you have a good understanding of the state-of-the-art and the various approaches that have been presented. For proposals submitted to mission agencies, the overall goal of the background section is to convince the reviewers that you have a complete understanding of the field, including work done by your colleagues and competitors. However, for proposals submitted to the NSF, the overall goal of the background section is not only to establish yourself as an expert, but also to educate the reviewer about the current state-of-the-art and recent developments.

7.4.3 Principle 3: Read the Call for Proposals!

This principle may seem obvious. Indeed, it is! However, in my proposal review experience, I've continually been amazed at how many proposal writers ignore vital information or proposal submission requirements clearly stated in the Call for Proposals, or other research opportunity announcement. Most research opportunity announcements will include restrictions on factors such as the following.

(1) Performance Period

The period of performance will be specified in the announcement, and will range from six months for SBIR Phase I proposals, to 10 years for large center NSF proposals. The typical performance period for a standard, single PI proposal will be three years. However, many funding opportunities will specify a one-year period of performance. Often the funding opportunities will permit follow-on proposals for additional funding. The follow-on opportunities are common for DOD funding agencies. For example, a typical DOD research grant will be for an initial three-year period of performance, with the possibility for additional research work supported through submission of a follow-on proposal for an additional three-year period. Proposals submitted to the NSF are typically funded for a three-year performance period. Generally, follow-on research is not directly possible, and any extension of the research will require a new proposal submission. The follow-on proposal can build upon the previous research results, but will need to clearly describe the work performed and accomplishments of the original research, as well as the research to be performed under the new grant. The new proposal should be submitted during an open window for proposal submissions, and will be evaluated and reviewed in competition with all other proposals received during the open period.

(2) **Budget Limits**

Most Call for Proposals or other research opportunity announcements will generally indicate the amount of funding that is available for the research opportunity, both for the overall program, and for individual proposals. It's very important to recognize these limits and to structure your research plan accordingly. If you fail to follow these guidelines, and make your budget either too small or too large, you will decrease your potential for successfully obtaining funding. An unrealistic budget submission will generally not prevent your proposal from being reviewed, and the budget is generally not a primary concern in the initial reading, unless the budget is completely unreasonable, or contains items not permitted under the guidelines. Indeed, the research to be pursued is the most important aspect of the proposal, and the reviewers' focus will be directed to this area. However, the budget will be considered in the evaluation, by both the reviewers and the program manager, and will be correlated with the proposed research effort. An unrealistic budget will be a negative factor in the evaluation. If the budget is too low, there will be questions regarding the scope of work and concerns if sufficient funds have been requested in order to permit the research to be successfully performed. Conversely, if the requested budget is too high for the funding available on the program, or for the research that is proposed, there will be questions concerning the breadth of the proposed research, the overall proposal focus, and questions regarding the ability to actually perform the proposed work with any realistic expectation for success.

If the proposed research is of high quality and the recommendation is to accept the proposal, the program manager may contact you and ask for a revised budget that is consistent with the available funding. In this case you will need to reduce the requested funding, and redefine the scope of the research and the specific research tasks to be pursued. When a budget needs to be reduced, quite often, and particularly for mission agencies, the program manager will contact you and inform you of the exact level of funding that will be provided. You will then need to revise your proposal accordingly. This is a common occurrence. For certain research opportunities, such as the NSF CAREER awards, there is a defined and specific budget amount that will be provided indicated in the announcement, and your proposal needs to be written for exactly this amount. The NSF will not proceed with the award of CAREER grants until the proposal budget satisfies the defined support level.

(3) Number of Pages and Font Size

Many research opportunities will indicate a limit to the number of pages that the proposal should contain. This limitation generally applies to the research narrative where the research that is to be pursued, along with the research plan and tasks, are described. For example, proposals submitted to the NSF are limited to 15 pages of research narrative. Similar page limits are generally associated with proposals submitted to other agencies and funding sources. The research narrative page limit does not include references, authors' biographies, budget, or any required statement and declarations, etc. The NSF 15-page research narrative limit must be respected, and the NSF FastLane system will not accept research narratives in excess of this limit. Often researchers wish to include more information than can be accommodated in 15 pages, and they will attempt to employ tactics such as the use of small font sizes, reducing page margins, etc. These tactics should not be employed. In particular, small font size makes it difficult for reviewers, and this could result in a reduced evaluation. Minimum font size is usually indicated in the announcement, and font size less than 10 point should never be used. Also, minimum page margins are indicated in the announcement and should be respected. You need to learn to prepare your proposal for the defined logistics, and the best approach is to concentrate on learning to present your ideas in a clear and effective manner, while respecting the proposal restrictions. Failure to do so could result in your proposal being returned to you without review.

(4) Address Every Scientific, Technical, or Programmatic Topic that is Described in the RFP

This is probably one of the most, if not the most, important principle for you to understand, and one which is often ignored. In fact, I've read many proposals where the Request for Proposal (RFP) clearly requested information on a range of topics that were described and outlined, sometimes in significant detail, only to have the author of the proposal pick-and-choose the items they wished to address. They would direct their proposal to certain topics on the list, generally the ones with which they were most familiar, and the ones to which their particular research was applicable, while ignoring the other topics. Although this approach is attractive for a single investigator, including only a portion of what is requested is a major mistake! When the RFP requests a list of topics to be addressed, the program managers have determined that they wish to fund a comprehensive research program to advance a given topic. They have already performed a strategic analysis of the topic and determined the advances that are necessary to more fully pursue the necessary research to achieve their overall objective. For example, in the program description they will often include a brief history and overview of the topic, including possible approaches, which have been determined from previous research. Often, in order to successfully compete for these funding opportunities, it will be necessary to identify and recruit researchers with whom you can collaborate, with each of the researchers focused upon a particular topic discussed in the RFP.

The program managers understand that it is unlikely that a single investigator will have the expertise to address all topics that are defined, and they expect that multiple investigators from various organizations will be included on the winning proposal. When the proposals are evaluated, the program managers will ask the reviewers to address all topics in the RFP, sometimes providing score sheets to be used, with each specific topic listed for evaluation and rating. If a proposal does not include that particular topic, it will be poorly rated. The main point is to read the RFP carefully, and make sure every requested topic is addressed. If it is not possible to address every topic, you should contact the program manager listed as the point-of-contact (POC), and discuss your research and possible participation in the research. The program manager may suggest that you participate with another group, or they may be willing to accept a limited proposal if your research is particularly attractive for one specific part of the program.

7.4.4 Principle 4: Make Sure Your Proposal is Not Summarily Rejected

One of the most frustrating and discouraging things that can happen is for you to identify a very interesting problem, develop a promising and novel research plan, and then spend many hours of focused effort writing an excellent proposal, and one that may also include results documented with data from previous research, and then you're ready to submit your proposal. You get your proposal submitted and approved through your institution's internal process, and you get the proposal submitted to the funding agency. Then you find that the agency won't accept your proposal, or that your proposal is immediately returned without review. How does this happen? And can you still get your proposal accepted for review by the agency. The answer to the second question is unfortunately "no," you're out of luck. This frustrating situation, although not common, does occur. So, what happened, and how can you avoid this situation?

The first thing you need to recognize and respect is the funding agency due date. Grant funding opportunities advertised on specialized RFPs, Calls for Proposals, Dear Colleague Letters, or other grant funding announcements, when the opportunity is directed towards a particular subject or topic, will almost always indicate a due date by which all proposals must be received at the funding agency. As a practical matter, there is essentially no or very limited flexibility on the due date and proposals received after the due date will not generally be accepted. This requirement is a hard date and is rarely extended, no matter the situation. In the days before electronic submissions, many proposal submitters would actually make trips to the funding agency to personally deliver their proposals and to make certain that the proposal was received by the funding agency before the due date.

As a personal experience, I recall one proposal I wrote many years ago and submitted to a grant funding agency. In order to have the proposal delivered to the funding agency by their due date I sent it through an express mail delivery organization, which guaranteed timely delivery. The proposal was scheduled to be delivered the day before the due date.

As you can probably guess, my proposal package was misplaced, and held up in delivery for a couple of days, which resulted in the proposal failing to be submitted by the funding agency deadline. The proposal was returned to me unread. When I checked with the express mail delivery service, they tracked the package, and then apologized for the late delivery. They then honored their guarantee by returning my mailing fee! Of course, this action left me out of the proposal competition. The electronic submission procedures now in place at most grant funding agencies prevent this sort of problem, but the due dates still must be honored and respected. As a principle, it is never a good idea to wait until the last minute to submit your proposal. There are many instances where computers crash, or where email, web-based systems, or networks may experience problems and be temporarily unavailable. Any of these problems could cause you to be delayed in submitting your proposal, particularly if you are planning to get everything submitted just a short time before the deadline. It is unlikely that any of these events would be accepted as excuses and result in the funding agency accepting your proposal beyond the due date. You don't want these types of failures to delay your proposal submission. Plan to have your proposal submitted with at least a day or two of lead time.

One exception to the necessity to meet the hard deadline for proposal submission is the occurrence of a natural disaster, where a certain degree of flexibility is permitted. The NSF, in particular, but other funding agencies as well, will permit delays in meeting the proposal submission due date. In the case of a natural disaster, proposers to the NSF need to contact the cognizant program director in the division or office to which they intend to submit their proposal, and request authorization to submit a "late proposal." If an adequate and convincing argument is presented, the NSF may permit an extension of the deadline by five business days. However, the extension must be approved in advance, and simply missing the published deadline by five days is not acceptable.

Other reasons that your proposal could fail to be accepted and returned to you without review relate to the failure to include mandatory statements, declarations, or other required information. Many grant funding opportunities will often request certain information be included in the proposal that is related to special topics not necessarily related to the research topic, or special requirements that are defined in the Call for Proposals, Dear Colleague Letters, or other research opportunity announcements. These special requirements need to be recognized and addressed. For example, the NSF requires that all research proposals submitted to it include a Data Management Plan (DMP). This is a supplementary document of no more than two pages labeled "Data Management Plan". This supplementary document should describe how the proposal will conform to NSF policy on the dissemination and sharing of research results. The DMP needs to be included, even if you do not anticipate that data will be generated in the course of performing the research. In this case, you would state in the DMP that you do not anticipate that your research will generate data that require management and/or sharing. Proposals that do not include a DMP will not be accepted and will not be able to be submitted on the NSF FastLane.

Another special requirement applies to proposals that contain post-doctoral researchers and are to be submitted to the NSF. These proposals must include a Mentoring Plan, which describes how a post-doctoral researcher engaged in the research will be mentored and advised, including a description of the mentoring activities that will be provided to them. Also, proposals that request funding for the support of post-doctoral researchers must include a Mentoring Plan as a supplementary document to the proposal. These requirements were mandated by Congress in Section 7008 of the America Competes Act.

Each of these items can be briefly described, usually in two pages or less. However, failure to include either the DMP or the Mentoring Plan will result in FastLane not accepting your proposal. That is, the FastLane system will not accept the proposal, and the proposal will not be distributed or even seen by a program director. Other funding agencies and sources may also include similar specialized topics that need to be addressed. For this reason, you need to learn to carefully read the research opportunity announcement and make note of any specialized requirement. In the proposal these topics need to be addressed in a very concise and effective manner.

7.5 The Basic Proposal

In this section we address the basic proposal structure and content, and describe how best to present the required information. I'll describe the main elements of a proposal from a generic perspective, while a proposal submitted to a specific funding agency may not require all the elements. However, all the elements discussed in this section will be required in proposals, depending upon the funding agency. For proposals submitted through the NSF FastLane or the US government Grants.gov websites, the various proposal elements are entered as separate documents, and it is very important to closely follow the submission format since the website will not accept documents that do not meet the requirements. More detailed information addressing the requirements for a specific agency is generally available on the agency website. For example, detailed information concerning proposal content required for proposals submitted to the NSF can be found in the NSF Grant Proposal Guide (http://www.nsf.gov/pubs/policydocs/pappguide/nsf15001/gpg_print.pdf).

There are differences in proposal organization and structure, depending upon the funding agency requirements and guidelines, but in general all research grant proposals will address and present information identifying and defining the problem that is being addressed, what new research is being proposed in order to contribute to advancing the topic and solving the problem, along with a plan for implementing and performing the research. Requested budget details, along with a description of who will participate in the research, are also presented, along with a description of the equipment and facilities available to be utilized in the research. This information is common to all proposals. Additional information address-ing specific requirements and details requested by particular funding agencies and sources may also be required, but this information and the relative details will vary from funding agency to agency. Often this information, such as letters of support and specific declarations and state-ments, will be included as an attachment to the proposal.

So what's included in a basic proposal? A basic research grant proposal, common to essentially all funding agencies, will include and consist of the following elements, in roughly the order presented below.

- A title page or cover page
- An abstract, executive summary, or project summary
- A table of contents
- The project narrative or project description

  - An introduction
  - A problem background section
  - Results from previous research
  - The research to be performed

- A statement of work
- References cited
- Personnel biographies
- The requested budget, including a budget justification
- A description of the laboratories, equipment, and facilities available to perform the research
- Special information and supplementary documentation

Each of these elements will now be briefly described.

7.5.1 The Title or Cover Page

Most funding agencies have specific requirements for the title or cover page, so you should make sure you follow them. Often, agencies will have standard cover sheet forms that will be completed by filling in the necessary information. For example, the NSF Fastlane system will request information that will automatically be transferred to the cover sheet, and this sheet will appear with your completed proposal. The title or cover page includes the title selected to briefly describe the research that is addressed in the proposal, along with the names of the principal investigator(s) who will perform and/or direct the research, the institution affiliation of the PI, including the department and university, the identification of the agency and address to which the proposal is being submitted, the performance period and dates of the research, the amount of funding being requested, and the signatures of the PI and the university official authorized to sign for the university. The date of the proposal submission is also often listed. The title should be brief, but accurately represent the research that is being proposed and should include keywords that relate to the research. The keywords are useful for database classification and searches. It is best to avoid overly lengthy titles. If more than one PI is involved, all names should be listed on the title page. However, only list those identified as principal investigators, and not others who will participate, but are not identified as a PI. The other contributors to the research will be included and identified in the personnel descriptions and budget. If PIs from more than one institution are involved, one institution should be identified as the lead, and the other institution should normally be indicated as a sub-awardee. If both institutions are on equal status and indicated as a lead, separate proposals from each institution are generally required. However, the separate proposals may be "linked" and reviewed together as a collaborative proposal. Before proceeding in this manner, it is best to communicate with the program officer managing the proposal submission process and clarify the collaborative approach. Collaborative proposals should be identified as such on the title or cover page. Often, agencies will encourage and have separate programs and procedures for reviewing and evaluating collaborative proposals.

7.5.2 Abstract, Executive Summary, or Project Summary

All grant proposals will include either an abstract, an executive summary, or a project summary, with the exact title depending upon the specific agency or organization to which the proposal is being submitted. The information included here provides the reader with a brief synopsis of what problem is being addressed, and the specific research to be pursued. This is a very important element in the proposal, equal in importance to the project description or narrative section, since the information provides the program officers and reviewers with their first impression of your research and what you propose to contribute to advance the topic. Also reviewers, when finalizing their review and evaluation, will often refer to the information presented here to remind themselves of the main contribution described in your proposal, so the information may also serve as their last impression of your proposed research. The abstract or summary should be carefully prepared, and generally only one page in length, and no more than two pages at most. The abstract should explain the key elements of your research project in the future tense in order to project what you propose to accomplish. That is, the proposed research is based upon established concepts, but you propose to advance the area by contributing new research, which is presented in the narrative or project description section and outlined in this section. Abstracts or executive summaries will state the significance of the problem that is being addressed, the specific goals and objectives of the proposed research, as well as how the research will be performed. The main point is to explicitly and briefly describe the problem that is being addressed, and clearly state what is new about the research that is being proposed. Often it is best to include explicit statements such as "The goal of this research is to investigate. . .," or "The objective of this research approach is to demonstrate a new. . . ." Many principal investigators will put these statements in bold type to make them obvious to the reader. Reviewers generally appreciate these enhanced statements.

The NSF requires that the proposals submitted to them include a project summary, rather than an abstract or executive summary. The NSF limits the project summary to one page. The proposal should include a section titled "Project Summary" that explicitly addresses two separate areas: (1) Intellectual Merit, and (2) Broader Impacts. It is very important to briefly and effectively include separate paragraphs addressing these two components since these areas serve as the basis for the proposal evaluation and are directly addressed by the reviewers. Intellectual Merit is generally not difficult to describe, since this is the main point of the proposal, and proposal writers generally do a good job in this area. The Intellectual Merit description contains the information generally included in a normal abstract or executive summary, and describes the significance of the research problem being pursued, and exactly what research contribution will be achieved. However, the Broader Impacts portion is more difficult to address, and many proposal writers either fail entirely to address the issue, or include an incomplete or sketchy description. While I've never seen a proposal receive an excellent rating and be accepted for funding based solely upon the Broader Impacts criterion, a good plan to address this area often makes the difference between acceptance and declination for high-quality proposals with equally rated Intellectual Merit descriptions. The Broader Impacts criterion must be addressed, and in a meaningful manner.

The Broader Impacts section does not contain information related to the performance of the proposed research, but rather to its significance and how it relates to other areas. The NSF Grant Proposal Guide clarifies what should be included in the Broader Impacts description through a series of questions intended to illustrate the criterion. For example: "How well does the activity advance discovery and understanding while promoting teaching, training, and learning?" "How well does the proposed activity broaden the participation of underrepresented groups (e.g., gender, ethnicity, disability, geographic, etc.)?" "To what extent will it enhance the infrastructure for research and education, such as facilities, instrumentation, networks, and partnerships?" "Will the results be disseminated broadly to enhance scientific and technological under-standing?" "What may be the benefits of the proposed activity to society?"

The NSF also states that "Broader Impacts may be accomplished through the research itself, through the activities that are directly related to specific research projects, or through activities that are supported by, but are complementary to the project. NSF values the advancement of scientific knowledge and activities that contribute to the achievement of societally relevant outcomes. Such outcomes include, but are not limited to: full participation of women, persons with disabilities, and underrepresented minorities in science, technology, engineering, and mathematics (STEM); improved STEM education and educator development at any level; increased public scientific literacy and public engagement with science and technology; improved well-being of individuals in society; development of a diverse, globally competitive STEM workforce; increased partnerships between academia, industry, and others; improved national security; increased economic competitiveness of the United States; and enhanced infrastructure for research and education." Many colleges and universities have established programs to address diversity and STEM education issues, and many successful PIs will coordinate and collaborate with these programs and describe their involvement in their NSF proposals. Also many successful PIs will recruit undergraduate students, as well as students from underrepresented minority groups to participate in the research they propose. The NSF encourages this approach and, in fact, offers additional financial support earmarked specifically for these activities. Once you receive an NSF grant, you should contact your program director and inquire about these enhancement opportunities, as they represent an excellent opportunity to expand your research activities.

When preparing the project summary in an NSF proposal, separate sub-sections should be written for and titled (1) Intellectual Merit, and (2) Broader Impacts. It's best to start each section with the words, the "Intellectual Merit of the proposed research is. . .", and the "Broader Impacts of the proposed research are. . .." Again, many PIs will place the words "Intellectual Merit" and "Broader Impacts" in bold type, which is a good idea.

7.5.3 Table of Contents

You may, or may not, be required to include a table of contents. However, it's a good idea to include a brief table of contents to indicate where major sections of the proposal are located. If you submit your proposal through the NSF FastLane system, the table of contents will be automatically generated by the system.

7.5.4 The Project Narrative, Project Description, or Statement of Work (SOW)

The project narrative, project description, or statement of work section is the main component of the proposal and is the place in the proposal where you provide an explanation of the research topic being pursued and where you define and present the new research that is being proposed. The exact title for this section will vary from funding agency to funding agency. For example, NSF calls this section the "Project Description," while most mission agencies, such as the DOD, DOE, NASA, NIH, etc., call this section the "Statement of Work." However, no matter the requested title for this section, essentially the same information needs to be provided. Information presented in this section lists and describes all essential and technical requirements for the effort to be performed, including standards to be used to determine whether the requirements have been met. This information is very important and constitutes the real substance of your proposal. In this section you will explain your research topic, why it is significant, what previous research has been performed, and exactly what new research you intend to perform, as well as how you plan to approach the project. This section is the heart of your proposal and needs to be very carefully written. Most reviewers will spend the majority of the time they devote to your proposal reading this section, along with the project summary or abstract. Many funding agencies will often place a limit on the number of pages that may be devoted to this section, so you need to very carefully explain your research in a concise and effective manner. For example, most mission agencies and other grant funding organizations will typically limit your proposal to no more than 12 to 15 pages, depending upon the funding organization. This page limit applies to the project description or statement of work section, and doesn't generally include information associated with references cited, contributor biographies, budget details, or other supplementary or facilities explanations and descriptions, etc. The NSF limits the project description section to no more than 15 pages.

The project description or statement of work section may require, and often will include, several subsections. In particular, for proposals submitted to the NSF, if you or your co-principal investigators (co-PIs) have previously been awarded NSF sponsored grants, or if the proposal is being submitted for additional support for a follow-on project, a synopsis of the previous results needs to be included in a clearly titled subsection. Other information typically included in separately titled subsections could include an introduction, a literature review, supporting or background information explaining the significance of the research problem, and a subsection devoted to describing the proposed research and related details, including a description of how the project will be performed. The latter material is the most important, and the other subsections are included in order to set the stage for the new research that is actually being proposed. Therefore, the preliminary subsections do not need to be lengthy, and only the material necessary to inform the readers and reviewers of what they actually need to know in order to understand what is actually being proposed, and the significance of the research, should be included. The main emphasis of the project description section should be directed towards explaining the proposed new research.

For proposals submitted to mission agencies, separate subsections may, or may not, be used and separately titled. Separate subsection titles, if placed on separate lines, take up precious space, so often the separate topics will simply be placed as separate paragraphs, or the title can be the first words of a new paragraph, with the subsection title words highlighted in bold type. However, for proposals submitted to the NSF, separately titled subsections should be used. As mentioned, if you or your co-principal investigators have previously had NSF sponsored research grants, or your proposal is being written for a follow-on grant for additional support to continue research on a particular topic or project, you need to include a synopsis of no more than five pages in length, which describes the research results you have already achieved on the previous research. This material should be placed in a separate subsection and clearly titled "Results from Previous Research." This synopsis, which is mandatory to include, applies to NSF grants received with a start date within the past five years from the date you are submitting the new proposal. This requirement applies to each PI and co-PI listed, regardless of whether the previous grant support was directly related to the research defined and described in the new proposal. If you or your co-PI(s) have received more than one NSF award during the five-year period, each PI need report only on the one award most closely related to the research described in the new proposal. The following information must be provided: (1) the NSF award number, amount, and period of support; (2) the title of the project; and (3) a summary of the results of the completed work, including accomplishments, supported by the award. The results must be separately described under two distinct headings, "Intellectual Merit" and "Broader Impacts." It's best to place these titles in bold type to make them readily apparent to the reviewer. You should also provide a list of all publications and presentations that resulted from the previous project. If the new proposal is for renewed support, a description of the relation of the completed work to the proposed work should be included. Reviewers will be asked to comment on the quality of the prior work described in this section of the proposal, so you want to briefly, but effectively, describe the previous research.

Since the project description section is limited to no more than 15 pages, the space devoted to the results from previous research will limit the pages you have to describe the new research you are proposing. The previous results material constitutes background for the research you are proposing, so you need to strike a careful balance between explaining what has already been accomplished, and what you are proposing to accomplish. Of course, you can limit your description of the results from the previous research in order to have more room to describe your proposed research. Reviewers generally are far more interested in the proposed research, so you should include only brief descriptions of the previous research, taking care to highlight important results, in order to have more space to include details that will enhance your description of the proposed project.

You should plan your project description, project narrative, or statement of work section carefully. Since the first sentences of this section serve as the first words that the reviewers will read, it is very important to briefly and effectively include an introduction to your research. The reviewers want a clear understanding of your research problem and topic. Therefore, the introduction should cover the key elements and points of your proposal. You should include a brief and clearly written statement of the problem that you are addressing, making sure to state and emphasize the significance of the problem. You should also indicate previous work on the problem, either your own research, or that performed by others. You should cite references to indicate the most significant advances of the past research that have been reported. This information provides background and rationale for the research you are proposing and establishes the need and relevance of the research and your approach. It is best to describe how your research differs from previous research on the same topic, but extends and advances the topic. You should also present a brief plan for how you propose to manage your project and perform the research. Finally, you should include some anticipated research results to indicate the nature of your proposed contributions. List only the principal goals or objectives of your research in the beginning of the project description and save sub-objectives for the latter part of the section. You want to indicate your problem and its significance at the beginning of the section so the reviewers will understand your problem, and then give them more details later, after they have had time to basically acclimate to your proposal.

You should include and provide an explanation for the timeline and expected performance period of your research project. In particular, indicate major milestones when critical advances or stages of the research will be initiated and completed. Reviewers will appreciate this information being provided in the form of a graphic, with major advances and items identified and listed on a vertical axis, and the performance period, usually time indicated in months, listed on the horizontal axis. Solid lines can be used to indicate expected periods of performance and related progress. This information can be briefly presented in this type of format, which presents a clear and visual presentation of the proposed timeline for your research project. This type of information can help reviewers understand your research goals, and provides them with confidence that you have clearly planned your research project and defined realistic milestones and goals.

One additional point relates to the inclusion of figures, graphs, diagrams, or tables. Inclusion of this type of graphics-based material is very important and enhances the presentation of the proposed research. This type of material can also enhance your ability to present effective arguments in an efficient manner. The old adage that "one figure is worth a thousand words" is definitely true, and inclusion of graphics-based data and visual information minimizes the number of words that are required to provide effective and focused arguments. In particular, graphics give the reviewer a clear understanding of trends, performance predictions, limitations, etc., and helps them understand the significance and importance of your proposed project. This type of information is particularly important for science and engineering proposals. You should try to put background information, projections, predictions, etc., in figures, graphs, or tables. Also, these can be located in the proposal using a multi-column format, which permits easy integration of the graphics and text, and helps you stay within the 15-page limit. The very best proposals that I have read made extensive and effective use of this type of presentation format.

7.5.5 References and Literature Review

You should include references to published research that is applicable to your proposed project. This information should be in a separate section titled "References," and all references should be numbered and identified in the text in the project description, project narrative, or statement of work section. The references section does not generally count in the page limit restriction, so there is little reason to minimize the number of references. You should carefully document the background material you include with references, which can refer to your own published research, or that published by others. However, only reference previous work that directly relates to your proposed project, either by supporting the significance of the problem you are addressing, or by providing information that relates to your proposed solution and approach. The reviewers will generally not carefully read this material, but want to see properly referenced proposals since the references provide an indication that you are fully aware of the most significant related work that has been published and reported and that you are fully aware of your research topic, the identity of your competitors, and what approaches they are pursuing.

Some funding agencies and organizations may require that proposals submitted to them include a literature review. The purpose of this requirement is that program officers and reviewers want to know whether you've done the necessary preliminary research to fully understand the subject you are addressing, and that you are completely aware of the state-of-the-art in the field and understand the main approaches and results that have been reported. This gives the reviewers confidence that you are prepared to perform the research presented in your proposal. A literature review, if required, should not be lengthy or exhaustive, but rather selective and critical, including the major publications and reports on the subject. Reviewers also want to see your evaluation of relevant research on projects that relate to your proposed research.

7.5.6 Personnel Biographies

The professional contributors who will work on the project should be identified in the personnel section. Everyone identified as a principal investigator (PI) should have a brief biography or CV of no more than a few pages included. The NSF limits the personnel biography to no more than two pages. Mission agencies will often not place a limit on the number of pages for the biography, but it is always best to keep it to less than five pages if a limit is not indicated. Overly long biographies can be distracting for reviewers, and counterproductive for the proposal author. In the biography the PI's current position should be listed, as well as a list of all the individual's previous academic and professional positions. The brief biography should also include a list of publications, presentations, patents, professional reports, etc. This will normally not include a complete list of the individual's publications, and should be truncated to include only the publications most closely related to the proposed research, with priority given to recent publications. For proposals submitted to the NSF, an additional list of a small number of examples that demonstrate the broader impact of the individual's professional and scholarly activities should be included. This list typically will include examples of the individual's efforts that were directed towards activities that address the broader impacts of the PI's research, such as the integration and transfer of knowledge, innovations in teaching and training, including the development of course materials and pedagogical methods, the development of research tools, the creation of computation methodologies, the development of databases to support research and education, and activities that broaden the participation of groups underrepresented in STEM. Also, service to the professional scientific and engineering community outside of the individual's immediate organization, including significant positions or offices held, should be listed. Examples include service as an Editor-in-Chief or Associate Editor for a professional journal, service as a Chair or organizer for a major professional conference or workshop, and other service that indicates professional stature.

Other personnel who will participate in, and contribute to, the research should also be identified, generally by name and position, but if no one has yet been identified or recruited, the position and purpose should be identified. A brief description of the staffing requirements and the details of what they will contribute to the research should be provided, along with a recruitment plan for new staff. These contributors will generally be listed as Other Personnel.

7.5.7 Budget and Budget Justification

A detailed budget that describes the expenses you determine to be required to perform the research must be prepared and included in the proposal. The budget is generally prepared for a year at a time, and a separate budget will be prepared for each year of the overall project period. For example, if a three-year project is being proposed, the budget will include separate budgets for each of the three-year periods, as well as a budget sheet that includes the totals for the entire project period. Permitted items that can be included in the budget are defined in the proposal guidelines published by the funding agency and generally also in the specific call for proposals or funding opportunity announcement. For proposals submitted to the NSF, information regarding allowable budget items can be found in the Chapter II of the NSF Grant Proposal Guide (reference listed in Section 7.5), and additional information can be found in the NSF Award and Administration Guide (https://nsf.gov/pubs/policydocs/pappguide/nsf16001/aag_index.jsp) regarding the allowability of the costs of certain budget items.

The budget will provide details for all expected project costs, usually listed in a spreadsheet or table, with each budget item listed as a line item. For proposals submitted to NSF through the FastLane website, budget items are entered into a spreadsheet template, and the correct budget format is automatically generated. The budget will be separated into direct and indirect costs (sometimes called overhead costs). The direct costs include the time committed to the project and salaries for all labor categories of personnel that will participate in the project, supplies that will be used on the project, travel to program reviews and conferences to present papers, travel to visit program managers or collaborators, costs to publish papers that describe progress achieved on the project research, tuition, benefits and fee expenses for students employed on the project, and any other item that is to be directly supported from the research funds. The indirect costs refer to charges that your institution collects for permitting the research to be performed. The items and services that colleges and universities can charge as indirect expenses are defined in the US government cost recovery principles described in the Office of Management and Budget Circular A-21 (https://www.whitehouse.gov/omb/circulars_a021_2004).

For research projects that are supported by US government funding agencies, the exact amount that can be charged for indirect costs is generally a percentage of the direct costs, a rate which is negotiated between your institution and the agency that is authorized to negotiate for the US government (called the cognizant agency). The majority of organizations will negotiate their indirect cost recovery rate (ICR) with the Federal agency that provides the preponderance of their funding. For the majority of colleges and universities the ICR is negotiated with the Division of Cost Allocation (DCA) in the Department of Health and Human Services (HHS), or the Office of Naval Research (ONR) Indirect Cost Branch, with each particular academic institution assigned to one or the other of these two organizations. These agencies review your institution's financial records on a periodic basis, and the indirect rates generally change slightly with each review. Your university research office or dean's office will make the current rates available to proposal writers, as well as provide a list of allowed charges for various budget items that are permitted to be charged to the research project. The total annual project budget will consist of a sum of the direct costs and the indirect costs, plus any equipment included in the budget. Generally, it is not permissible to charge indirect costs on equipment purchases, so these charges are listed as a separate charge after the indirect costs have been calculated and listed. For multi-year proposals, approved inflation increases are generally included for the second and third years of the project budget. Your research administration office or your dean's office will generally have this information available.

The budget should also include a narrative, or description, of how the requested budget funds will be used. In particular, the budget narrative should justify the need for the various budget items, and provide an explanation of how the various items are important for the performance of the research. The Call for Proposals, or other grant funding opportunity announcement, may, or may not, specifically request a budget justification. However, even if the proposal guidelines do not specifically mention a budget justification narrative, you should include a brief explanation of the budget. Generally, only a one- or two-page budget justification is satisfactory. The NSF, in fact, limits the budget justification to no more than three pages. Your budget justification should be brief, and clearly state the need for various budget items. Justification should be provided, in particular, for any equipment or instrumentation items you propose to purchase, or travel you plan associated with the research. Travel to conferences or other visits for reasons associated with the performance of the research is permitted, but needs to be explained and justified. Also, many agencies require details concerning travel to conferences at the time of proposal submission, and this requirement can be a challenge since conference details are often not available a year or more in advance. For these trips you will need to identify the conference, and then estimate the expenses. It is best to include such expected travel in the budget, described in generic terms, since it is difficult, often requiring agency approval, to modify the budget to include the travel after a project has commenced.

7.5.8 Current and Pending Support

Funding agencies require a listing of all current and pending research grants and financial support that you and your co-PIs have been awarded for the five-year period preceding the date you submit your proposal. In this list you should also include the current proposal that is being submitted. There are basically two reasons for this requirement: First, funding agencies want to ensure that the major participants identified in the proposal are not overcommitted and that the sum of their research effort commitments, including the proposed research, does not exceed 100%; and second, to determine that the proposed research project does not overlap with research supported on other projects or funded by other grant funding sources. That is, the sum of all the current and pending effort commitments, expressed as a percentage, for all the major participants in the proposed research project may not exceed 100%. Also, the same research project cannot be separately funded by different agencies, unless they each agree to partly fund portions of the research. However, the funding agency program managers need to be informed of all sources of funding provided to the project. All current and continuing research project financial support that you and your co-PIs have been awarded, even if you receive no salary support from the listed projects, from all sources, including Federal, State, local or foreign government agencies, public or private foundations, and industrial or other commercial organizations, must be listed. Information that should be listed includes the project title, the dates of performance, the number of person-months per year devoted to the project, and the identity of the funding agency.

The same proposal may be submitted to separate funding agencies, but the proposal should be separately listed for each agency to which it has been submitted. Also, each funding agency needs to be informed of the duplicate submission. Funding agencies will include a section for identifying other agencies to which the proposal has been submitted. Submission of the same proposal to multiple agencies is not something that should necessarily be avoided. For example, if one, or both, of the funding agencies decide to fund the proposal, they will either ask that the proposal be withdrawn from the other agency, or they may, at their discretion, contact the other agency, and if both want to fund the research, they may agree to co-fund the research, with each agency providing a portion of the requested funds. However, in this case, there will be no increase in the amount of funding provided, and the PI will need to report to both agencies.

7.5.9 Facilities, Equipment, and Other Resources

Funding agency program managers, as well as reviewers, want to verify that the research presented in the proposal can actually be performed. Particularly for experimental research that may require access to laboratories, equipment, or instrumentation not generally available, this can be a major issue. Therefore, information related to the availability and adequacy of all necessary laboratory facilities and other resources that can be used to perform the proposed research needs to be identified, listed, and explained. If the research is to be performed in existing laboratories, equipped with the equipment and instrumentation necessary to perform the research, the laboratory facilities should be described. Often universities will operate and maintain common laboratories that are available to all faculty members and their students. Normally, a fee is charged for access and use of these facilities. The available equipment and instrumentation related to the proposed research project should be identified and described, along with the access rules, including the fee structure and operating rules. If safety training is required, this should also be explained. Basically, the proposal writers should include a brief, but complete, aggregated description of the internal and external facilities and resources, both physical and personnel, that will be available to the PIs and their students to perform the research. If the proposed research primarily requires computer resources, the available computing facilities should be described. Often, only PC-based computing will be required, but if access to supercomputer facilities is required, how this access is to be acquired should be explained, whether the facilities are local, or if access through a US government facility or other organization is planned.

7.6 The NSF CAREER Proposal

The NSF Faculty Early Career Development (CAREER) Program is a special program designed for, and directed towards, support of the development of young faculty members early in their careers. As stated by NSF, the "CAREER Program is a Foundation-wide activity that offers the National Science Foundation's most prestigious awards in support of junior faculty who exemplify the role of teacher-scholars through outstanding research, excellent education and the integration of education and research within the context of the mission of their organizations." The CAREER Program is directed towards assisting young faculty members in building a firm foundation that will assist them in establishing a lifetime of leadership in integrating education and research. The CAREER Program is not a research program in the typical sense, but rather an integrated program consisting of both research and education, with the overall goal of helping the faculty member establish a life-long professional plan. The CAREER grant is awarded for a period of five years, with a dollar amount of a minimum of $400 000 distributed over the five-year period. The actual amount varies by NSF directorate, so you need to check with the particular directorate to which you plan to submit your proposal to determine the amount. You should plan your budget to be exactly the amount that the directorate provides, as a budget over the defined amount can result in your proposal being rejected, and an amount that is less can cause delays, and the need to modify your budget until it's in compliance with the permitted amount. More information regarding the NSF CAREER Program can be found on the NSF website (http://www.nsf.gov/career).

A CAREER award is considered very prestigious by both the NSF and the academic community and, as a result, the program is highly competitive. Many universities will routinely boast of the number of their faculty members who have received CAREER awards, and will advertise and distribute information related to their CAREER awardees. For a young faculty member the receipt of a CAREER award is a major achievement and is highly valued. It is also a great accomplishment and asset that is considered as a very positive factor in the promotion and tenure process. Therefore, a very large number of new faculty will submit CAREER proposals. Unfortunately, the large number of proposals in any given competition, coupled with a slowly increasing NSF budget, results in fairly low success rates, which vary by directorate, but typically range between 10% and 20% for the CAREER Program.

Young faculty members generally do not fully understand the differences between a regular research proposal and a CAREER grant proposal, and focus their proposal primarily upon their proposed research activities. This is a major mistake, and often results in the proposal being declined. In this section, I'll explain the NSF CAREER grant proposal in some depth, focusing upon what needs to be included.

7.6.1 Differences Between an NSF CAREER Proposal and a Regular Research Grant Proposal

The major difference between an NSF CAREER grant and a regular research grant proposal is that the CAREER proposal is not a research project proposal, but rather is a career development award. Your proposal must reflect this focus. The proposal should describe a path to a future career, not a specific research project. You need to determine your research path in terms of your lifelong research goals, and then identify milestones to reach your goals. The first one or two of these goals will serve as the research projects for your CAREER proposal. The research goals will involve the integration of research and education activities, and although the overall goal is a lifelong plan, the milestones and integrated research and education plan need to be defined and structured for the five-year period of the NSF grant in order to establish the viability of the plan. However, the overall goals should project into the future.

All CAREER proposals must be structured with an integrated research and education plan. Although it is understood by NSF program directors and the proposal reviewers that there is no single or best approach to an integrated research and education plan, they want you to think creatively about how your research will interact with your education activities.

They want to see creative approaches and plans that will effectively couple research with education. There are, of course, different expectations within various disciplinary fields and organizations, and a wide range of research and education activities may be appropriate for the CAREER program. When planning your activities, you should address three basic questions: (1) Where are you today, and what have you accomplished? (2) Where do you see your future and what do you want to accomplish in five, 10, or 20 years? (3) How do you plan to get from where you are today to where you want to be in the future? In formulating your career plan you should consider your expertise and interests, your career goals, and your position and the resources that are available to you. Your CAREER proposal should be consistent with your goals. Your CAREER proposal should also be compatible with your home institution's goals, and your CAREER plan should represent a contribution to society at large.

Structurally, the CAREER Program grant proposal has the same major elements discussed in Section 7.5, and the proposal will be submitted through either the NSF FastLane or Grants.gov websites, and must follow the guidelines presented in the NSF Grant Proposal Guide (GPG). If Grants.gov is used, the same basic elements as required for NSF FastLane submission will be used, as discussed in the NSF Grants. gov Application Guide, which is available on the Grants.gov website or the NSF website (http://www.nsf.gov/publications/pub_summ.jsp? ods_key=grantsgovguide).

The basic elements of the proposal are as follows.

- A cover sheet
- Project summary
- Table of contents
- The project description
- References cited
- Biographical sketch of the principal investigator
- Budget and budget justification
- Current and pending support
- Facilities, equipment, and other resources

- Additional supplementary documentation required for CAREER proposals

Each of these elements is separately prepared and entered into the website template, and the website will automatically generate the complete proposal in the NSF proposal format.

Major differences between the CAREER proposal and a regular research proposal are primarily in the PI eligibility, the proposal cover sheet, the project summary, the project description, and additional supplementary documentation required for CAREER proposals. The cover sheet must contain the word "CAREER" in the proposal title, followed by the descriptive title of the proposal. The descriptive title should briefly, but concisely and accurately, represent the substance of the contents of the proposed research.

The other differences are discussed in the following subsections.

7.6.2 Eligibility

Junior faculty members at all CAREER-eligible organizations are permitted to submit CAREER grant proposals. The term "junior" faculty member refers to both tenure-track Assistant Professors, and those in an equivalent rank. There is no citizenship requirement. Proposers must hold a doctoral degree by the proposal submission date, and be employed in an untenured position at the time of proposal submission. They must continue to be untenured until at least October 1 following the submission date. They must be employed at an organization located in the USA, its territories or possessions, or Puerto Rico, that awards degrees in a science, engineering, or education field supported by the NSF. Faculty members at the rank of Associate Professor, with or without tenure, are not eligible to compete for CAREER grant awards. Scientists and engineers employed by non-profit and non-degree-granting organizations such as museums, observatories, or research laboratories are also eligible to compete for CAREER grants, providing they satisfy the other eligibility requirements. The NSF especially encourages CAREER grant proposals from women, members of underrepresented minority groups, and persons with disabilities. A PI may submit only one CAREER grant proposal in any given year competition, and they may submit proposals in up to only three CAREER grant competitions. After the third CAREER grant proposal submission, a PI is no longer eligible for any future CAREER grant competitions, even if none of the three proposals that has been submitted in previous competitions was successful. A PI may receive only one CAREER grant in their lifetime, no matter how many times they change employment locations. However, it is permissible to transfer a CAREER award from one institution to another if the PI changes institutions. In this situation, both institutions, as well as the NSF, must concur and approve the transfer. There is no limit on the number of proposals that can be submitted from a given institution.

7.6.3 The Project Summary

The project summary should be very carefully written and structured. The NSF limits the project summary to one page, and this restriction must be respected. You should write your project summary in basically four paragraphs. The first paragraph should address your research, and the paragraph should contain statements that briefly and clearly state your long-term research goal, the research objective of your CAREER proposal, and a description of the approach you intend to follow. The second paragraph should address your education plans. You should include declarative sentences that state your long-term education goal, the education objective of your CAREER proposal, and a description of the approach you propose to follow. The third and fourth paragraphs will present the required statements addressing Intellectual Merit and Broader Impacts. Those paragraphs should begin with the words Intellectual Merit and Broader Impacts, and it's best to highlight these words using bold type. Experienced reviewers expect to see the project summary information in the described format, and using any other approach will likely result in a degraded rating for your proposal.

7.6.4 The CAREER Grant Proposal Project Description

The project description section is the main element of the proposal, just as it is for a regular research grant. However, there is a fundamental difference in the content and approach described in the project description for a CAREER grant and for a regular research grant. In particular, the CAREER grant project description needs to contain a complete and detailed description of your long term *career* plan, including both your research and education objectives. The description needs to be carefully written, particularly since it needs to describe both the research and education plans, as well as set out how they are to be integrated. Just as for regular research grant proposals, the project description section is limited to no more than 15 pages. The project description should contain a well-developed and detailed plan describing activities that will, over the five-year project period, establish a foundation for a lifetime of contributions to research and education. The project description should address four major areas, consisting of: (1) a description and plan for your proposed research project, including preliminary supporting data and results, where appropriate, and the detailed plan to achieve the overall goals, along with the expected significance of the anticipated results; (2) a detailed description of your proposed educational activities, including plans to evaluate their impact on students and other participants; (3) a description of the integration plan detailing and explaining how the research and educational activities are integrated with one another; and (4) a synopsis of the results from previous NSF grants, if applicable. The previous results synopsis is limited to no more than five pages, just as for regular NSF research grant proposals. Also, it is not necessary to include separate descriptions of research and educational activities if they are inter-related in such a manner that the overall program can be described as a structurally integrated and interdependent program.

Your proposed research should be original and directed towards a significant problem. Projects that address incremental advances should be avoided, and proposals that contain only incremental approaches are not generally successful. Research is a process of discovery where there is a structured and determined effort to learn something that is not already known. Scientific, engineering, and educational research is always an extension of an established and known knowledge base, and the scientific method is orderly, repeatable, and verifiable. You should state your research objective clearly, and in a manner that leads the reader directly to the conclusion that your approach will lead to the desired result. The best research proposals will begin with a clear and concise statement of the problem that is being addressed, and then provide a brief description of the state-of-the-art, including references to document major and significant advances that have been reported. You should address four critical questions that experienced reviewers will have in mind and will be expecting you to answer. First, the reviewers want to see a clear explanation of what your proposal is addressing. That is, what research problem are you addressing, and why is it significant? For CAREER proposals, as already stated, you need to include clear and concise statements of both your research and educational objectives. Second, the reviewers need to have confidence that your approach to your proposed research and educational objectives will be successful. You want to make sure the approach that you present in your proposal is realistic, and leads naturally from the statement of the problem to the desired goal of logical and successful advancement in the research topic area. They also want to know that you have the necessary laboratory facilities, equipment and instrumentation, computing access, and other resources necessary to perform your proposed research. Third, the reviewers need to have confidence that you are prepared to perform the proposed research project. In particular, they want to see some preliminary results or data that give an indication that you are on the correct path, or other background information that supports your approach. Fourth, you want to help the reviewers form an opinion of the overall value of your proposed research, and the scientific contribution that you will be pursuing. Basically, they will have the fundamental question: "Is the research worth pursuing?" You should address this question in the Broader Impact statement.

Your proposed education plan can involve a very diverse and broad range of activities, and should be directed towards the involvement of other groups of people on levels ranging from K-12 students, high school teachers, undergraduates, graduate students, and the general public. You need to have an outreach plan that will engage one or more of these groups. Typical activities could include involving other people in your research using new tools, laboratory methods, novel learning techniques, and other approaches. These activities should be related to your proposed research, and should be directed towards outreach and involving other people in your research. You want to be able to bring the excitement of your research to help, and hopefully inspire, the education of others.

You also want to seek new methods to deliver your research results to a broader audience than those in your immediate research community. Often, many colleges and universities have established active outreach programs that routinely recruit and engage people from secondary schools, college undergraduates, students from minority and underrepresented groups, students with disabilities, etc., in campus educational and research activities. These programs will often include activities such as summer research camps for K-12 students, secondary teacher camps and seminars, teacher interns helping in your research program, and other similar activities. Many CAREER applicants will direct their educational activities to these programs, and integrate their program with that of their college or university. This type of leverage is encouraged and represents an excellent method to develop an effective education program that naturally integrates with their research activities. Your education program should be both innovative and creative. Finally, you should also include a plan for evaluating the effectiveness of your integrated education and research program. The development of a set of practical metrics for this purpose would be considered a plus. A truly novel and innovative education program, effectively integrated with your research program, will be very positively received by the reviewers.

7.6.5 Additional Supplementary Documentation

For the CAREER program, the NSF requires that your department head or department chair provide a support letter stating institutional commitment to support you as you pursue the career development plan described in your CAREER proposal. The letter can be no longer than two pages. If you have appointments in more than one department, the head or chair of the department that will grant you tenure should sign the letter. However, if tenure is to be granted in more than one department, the letter must be signed jointly by the department head or chair of each department. Only one support letter will be accepted. This support letter is mandatory, and proposals submitted without this letter will be returned without review. The letter needs to commit to provide institution support for the PI's proposed research and education activities. The letter also needs to provide a description of how the PI's career goals and responsibilities are consistent with the department's goals and priorities. Your department head or chair must also commit to support your professional development with mentoring, and provide support and resources that may be necessary in order to assist you as you implement your research and education plans, along with their integration. The letter also needs to explicitly state that you are eligible to participate in the CAREER program.

Letters that indicate collaboration are permitted to be submitted, but are not mandatory, and proposals without these letters will be accepted. The letter of collaboration would be written by other scientists, engineers, or professionals, generally from other organizations outside your institution, and with whom you intend to collaborate as you pursue your CAREER project. The letter of collaboration is limited to stating the intent to collaborate. Letters of recommendation from collaborators are not permitted, and the collaborator should not provide an opinion of the quality of the proposed work, the quality of the PI, or comment otherwise upon the proposed project. Also, a letter of collaboration should be in a single-sentence format. That is, the letter should simply state that, if the PI's CAREER project is selected for funding, the author of the letter intends to collaborate on the project, or commit resources as detailed in the project description. If collaboration is planned, the details, including an explanation of the need for, and nature of, the collaboration and the intellectual contribution the collaboration will bring to the project should be described in the project description. Also, permission to use a collaborator's facility, including laboratory, equipment, instrumentation, and any offer by the collaborator to provide samples and materials for research, logistical support to the research and education program, or mentoring of students, should be described in the project description. The collaborator's position and organization should be indicated, but nothing else should be in the letter. Letters from multiple collaborators are permitted.

7.6.6 The PECASE Award

The Presidential Early Career Award for Scientists and Engineers (PECASE) is an honorary program, and is considered to be the highest honor the US government confers upon outstanding scientists and engineers in the early stages of their independent research careers. The agencies participating in the PECASE award program are: the Department of Agriculture, the Department of Commerce, the Department of Defense, the Department of Education, the Department of Energy, the Department of Health and Human Services, the Department of the Interior, the Department of Veterans Affairs, the Environmental Protection Agency, the Intelligence Community, the National Aeronautics and Space Administration, the National Science Foundation, and the Smithsonian Institution. The number of PECASE awards for each agency is proportional to their basic research budget. PECASE awardees must be employed by or funded by one of the participating agencies, and they must be US citizens or permanent residents and have received their PhD degree within five years of the nomination. The NSF selects its PECASE nominees each year from the CAREER awards that they consider the most innovative, creative, and meritorious. Selection of the PECASE award nominations is based upon two criteria: (1) the innovation of the research, which is considered to be at the frontiers of science and technology relevant to the NSF mission; and (2) the involvement in community service of the CAREER grantee, as demonstrated through scientific leadership and community outreach. The CAREER awardee does not apply for a PECASE award, and the nominations are selected by the NSF. There is no monetary award involved with the PECASE program. The final selection and announcement of the PECASE awardees is made by the White House Office of Science and Technology Policy.

7.6.7 Common Mistakes

There are several common mistakes that many PIs make in preparing their proposals, and these mistakes can degrade the ratings that they receive and result in their proposal being declined. These mistakes are easily avoided. The most common mistakes, as identified and stated by reviewers, are as follows.

- Ignoring the rules presented in the NSF Grant Proposal Guide. If these rules aren't followed, your proposal has a high probability of being returned without review. The NSF is not lenient on the failure to follow the rules.
- Planning the proposed research and education on too broad a basis.
- Planning the proposed research and education on too narrow a basis.
- Basing your proposed research upon an incremental advance.
- The proposed research plan is not likely to yield results that will successfully meet the goals of the project.
- The research project methodology and design are flawed and not well designed.
- The resources needed to perform the research are either not available or are inadequate to perform the research described in the proposal.
- Submitting an unrealistic budget by making it either too large, or too small. The budget needs to directly correlate with the work described in the proposal.
- Focusing the research program on development efforts, computer programming, commercialization, or design. Where possible, avoid words such as "develop," "design," "optimize," or any other word that distracts from "research." The proposed work needs to be focused upon fundamental research and scientific discovery.
- Failure to include an adequate education plan, or presenting an education plan that is generic, and proposing to do what is normally expected of any PI in your field.
- Failure to demonstrate knowledge of education problems or to demonstrate understanding of what is effective in education.
- Failure to provide a realistic integrated research and education program. Also, paying inadequate attention to the education program component of your career plan.

- Failure to include outreach or engage with K-12 students, undergraduate students, or students from underrepresented minority groups.

7.7 What to do if Your Proposal is Declined

Let's assume the worst case scenario. You've followed all the proposal preparation guidelines, sought and received advice from experienced proposal writers, and prepared and submitted what you consider to be an excellent proposal, only to be informed that your proposal has been declined. What are your options, and what should you do? First and most important, don't become frustrated or disillusioned. Having a proposal declined is always disappointing, but it happens to everyone, including the very best researchers. You want to learn from this experience and be better prepared for your next proposal submission. Therefore, you want to know the reasons for the decision to decline your proposal, and find out what deficiencies were identified, and what you need to do to correct them. If you submitted your proposal to the NSF, the summary of the panel review is available to you on FastLane. However, the panel summaries generally contain only a brief, and sometimes unsatisfactory, synopsis of the panel discussion. The summaries generally do not include detailed critiques, or indicate in detail how the proposal can be improved. Comments addressing these issues will be in the summaries, but, in general, they will be cursory. If your proposal was submitted to a mission agency, you may have received minimal information regarding the proposal review, and possibly have simply been informed of its declination.

Your first step should be to contact the program director or program manager who was responsible for reviewing your proposal. If you submitted your proposal to a mission agency, the program manager is, most likely, the person to whom you sent your proposal. If the proposal was submitted to the NSF, it was assigned to a program director, and he or she selected the reviewers that participated on the proposal review panel. The NSF program directors are organized in their respective divisions according to subject and topic area, and the program director you should contact is the person who has responsibility for your research subject, and the program director who communicated with you regarding your proposal. When you contact them, you should request a debriefing on the reviews and evaluations your proposal received. They will provide you with information regarding the discussions relative to your proposal. However, it is not likely that you will receive a significant amount of information regarding the panel or reviewer evaluations, and you won't be given any information that could reveal the reviewers' identities. The debrief information generally provided to proposal writers tends to be fairly general, although occasionally specific critiques will be provided, particularly if some obvious and glaring deficiencies were noted in the proposal. This type of information is easy for the program manager or program director to report. In fact, this is the main information you are seeking. You want to know what the reviewers considered to be any major weakness or deficiency, and particularly if the same weaknesses were noted by the majority of the panel members, or only one or two reviewers. This is the most important information that will help you make improvements for your next proposal and, in particular, you will want to address and correct deficiencies or weaknesses noted by multiple reviewers.

Proposals will naturally fall into three groups; excellent proposals, weakly developed or poorly written proposals, and those in between. The first group will be rated highly by the majority of reviewers, and these proposals will be kept in the competition. The second group will be lowly rated by most reviewers, and will essentially be eliminated from the competition. The third group is the most difficult to review, and often will receive mixed evaluations, ranging from fairly high to fairly low, with most ratings somewhere in the middle. The review panel will discuss the middle group of proposals in some detail, and will end up with an overall evaluation and rating for each proposal. The issue is that sometimes mistakes are made, particularly by reviewers with a general knowledge of the research area, but without intimate knowledge of the proposal subject area. This is the reason that you would want to know if weaknesses were noted by multiple reviewers. If a weakness is stated on a concept presented in the proposal, and particularly if you do not agree with the reviewer, you may very well be correct and, in this situation, you should definitely not make any changes. Reviewers serving on proposal review panels have been known to make mistakes. For example, I've served on several proposal review panels where I've heard a particular reviewer say something like "This work is not new, it's already been done and reported," when, in fact, the work described in the proposal has not been done, and the research is expanding a very important area. These types of mistakes sometimes are not caught during the panel review process, and can result in a proposal declination. In this situation you do not want to make any changes since a proposal submitted to future competitions will be reviewed by different reviewers.

You need to keep in mind that there is always a limited budget for any particular grant funding competition, and only a limited number of proposals can be selected and funded. In fact, most competitions receive a large number of proposals and the rivalry can be intense. This results in many proposals that receive high ratings not being selected due to a lack of available funding. However, funding agencies require that reasons be stated for each proposal selected, as well as for each proposal that is declined. To satisfy this requirement, there will be statements entered into the proposal evaluation for highly rated, but not selected, proposals to justify the declination. These statements do not generally provide information that will be useful for future proposals.

For proposals that are submitted to mission agencies and declined, the debriefing by the program manager provides an excellent opportunity to learn more about his or her particular program and what research topics are being supported. Your proposal may have been declined since the program manager did not feel the subject effectively fit into his or her particular program. Your proposal may have been, in fact, high quality and worthy of support, but was not selected due to lack of interest by the program manager. You want to learn as much as possible about the program manager's research program and try to align your proposed research to topics in which they have interests.

7.8 What We've Learned

In this chapter we've discussed issues related to proposal preparation, submission, and the review process. We've also discussed how proposals are reviewed within various grant funding agencies, and noted that some agencies will send proposals to selected reviewers where the proposal will be independently reviewed, while some agencies make almost exclusive use of review panels. However, all proposals will receive multiple evaluations, and will be reviewed by three to five reviewers, and sometimes more. We're learned that proposal reviewers are your colleagues and all are professionals and highly trained in their respective fields. Experienced reviewers learn how to read and evaluate a proposal quickly and efficiently, and they will look for certain critical information. This information is stated in four fundamental principles, that should be followed; know your subject, learn how to write your proposal for the intended funding agency and its reviewers, make sure to read the details of the Call for Proposals, or other grant funding announcement, and include all requested information, and follow all stated procedures so that your proposal is not summarily rejected.

The components of a basic proposal were defined and discussed. The importance of preparing a brief and effective abstract, executive summary, or project summary, in which you present a description of your proposed research and your research plan was indicated. The main component of the proposal is the project narrative, the project description, or the statement of work section. The NSF CAREER proposal was separately discussed, since this program differs in a major way from a normal research grant proposal. That is, the NSF CAREER program is not solely a research program, but is a long-term career development program, and it requires an integrated research and education program. Finally we discussed your options in the event your proposal is declined. In particular, it's very important to contact the program manager or program director to learn the issues that resulted in the declination of your proposal. In this manner, you want to learn how to correct any deficiencies and address any weaknesses in order to improve your prospects for future success.

8   Cautions and Other Concerns

There are several requirements and restrictions that you need to understand regarding how you conduct your research and how you present your results to the general community. Operating in an academic environment offers a significant degree of freedom and flexibility in the projects in which you choose to become involved, and academic life can be very rewarding. Indeed, the freedom and the ability to pursue your own research interests and to build a successful research and education program are the primary reasons many faculty members are attracted to an academic career. However, along with the freedom there are various institution and government policy guidelines that are used to ensure that resources are properly utilized. Some of the guidelines affect operating procedures, and some are legal restrictions that define the type of activity that can be conducted. In this chapter we'll examine some of the major policies that govern the performance of research in an academic environment.

8.1 Disclosure Requirements

Disclosure policies require that you inform your institution of any possible involvement with outside interests that you may have established. For example, typical involvement with outside interests could include income you receive, usually above some limit such as $10 000, from any external company or organization that does business with the university, sponsors research or other projects with the university, or hires students or other employees that you supervise, or for whom you are otherwise responsible. The income could consist of salary, consulting fees, honoraria, royalty payments, the proceeds from the sale of any equity you may own, or any other mechanism that provides you with financial income or resources. You also need to disclose any investments or equity you own in an organization or entity that does business with your institution, or that provides support for projects or hires students. The investments could include stocks, bonds, or other equity instruments above a certain monetary value. Any position you might hold with an organization or entity that does business with the university, sponsors research or other projects, or recruits students also needs to be disclosed.

Also, it's becoming more common for some faculty members to be involved with start-up businesses or other entrepreneurial activities, and they often serve as founder, chief executive officer, chief technical officer, partner, director, manager, or some other position within the outside organization. In addition, many faculty members serve as paid consultants to various business and government organizations. Any activity of this type needs to be disclosed. Also, if you author a textbook, or other classroom materials that will be sold commercially, and use this material in a class in which you are the instructor of record, you must disclose this information to your institution. None of these activities is forbidden, but university policies require that all external engagements and activities be disclosed and approved, and that, for certain activities, a management plan be prepared and approved before engaging in the outside activities. The university management will generally work with you and assist you in developing a management plan so that the outside affairs can be effectively controlled and monitored. Engagement in the outside activities will be approved by your institution, generally by your department head or chair. If you plan any outside activities, you should discuss your plans with your department head or chair, or your college dean, particularly if the activity involves participating in a start-up business, which will generally require you to be released from some of your academic responsibilities for a period of time. Academic institutions have very well-developed policies and procedures that satisfy all local, state, and national laws that govern faculty engaging in external activities, and the university guidelines and procedures will ensure that all required policies are followed. The disclosed information will be provided to funding agencies to which you submit proposals, in accordance with their conflict-of-interest disclosure policies.

As you write and submit proposals to funding agencies, you may wish to submit a given proposal to more than one funding agency in an attempt to maximize your chances for receiving funding. Multiple submissions are permitted, but you must disclose to each agency to which the proposal is submitted all other agencies to which the proposal has also been submitted. This requirement is very important and should not be neglected. It is not permitted to accept funding for the same work described in a proposal from multiple funding agencies. In fact, to do so is considered fraud, and penalties can range from being denied the ability to submit future proposals for some period of time, to financial penalties, to imprisonment in extreme cases. It's best to make sure that the research you propose to any given funding agency is distinctly different from that you describe in separate proposals to the same or different agency, even if the research is similar in nature, and the projects are related to each other. For proposals simultaneously submitted to two agencies for review and evaluation, if one of the agencies that reviews your proposal decides to proceed with an award, they will contact you and request that you withdraw the proposal from the other agency. If both agencies decide to proceed with an award, they may agree to co-fund the project, with each agency providing a portion of the total funds. This is an excellent method for program managers to leverage their research budgets and increase the number of projects they are able to support.

A similar principle applies to follow-on proposals to extend research performed in a previous grant. The new work must be distinctly different from the previous research, although the research will be similar. In fact, research projects that relate to each other are encouraged, and can extend the range of research that is performed, permitting more progress to be achieved, more results to be obtained, and more students to be employed. However, each project needs to be separately focused and described. For the new proposal, results obtained on the prior project will naturally lead to new proposed research, particularly if the research is in an area of current interest, and the overall project has a long-term goal or objective. A description of the previous work, and the presentation of the results that have been obtained, provides strong evidence for the new proposed approach and helps to strengthen the case for renewed or additional funding. However, the research presented in the new proposal should be a clear enhancement and extension of the previous work. The previous project should be clearly identified and referenced in the new proposal. Program managers and program directors view this approach favorably.

8.2 Summer Salary Limitations and Time Commitments

At most US academic institutions faculty members are paid on a partial year basis. Generally, they receive salary for the academic year, which is usually nine months, although the exact number of paid months varies, depending upon the specific college or university. The academic year salary is paid over the 12-month calendar year so that the faculty member receives a paycheck every month. However, the three-month summer period is not actually a period included in the salary schedule. Faculty members are permitted to engage in paid research, teaching, or other activities during the summer months, but they need to secure the funding for their summer salary on their own initiative. Often, the college or university will offer summer classes that the faculty member may teach, and the institution will pay them additional salary for their effort. However, most faculty members engaged in research will devote their summer months to focused and directed research activities. They will receive salary from research grants that they have obtained, or other funding that they have secured. At the majority of research institutions they are permitted to charge up to three months of the summer period to work on research projects, although some institutions will limit summer salary to two months. Of course, the summer salary needs to be included in proposals that they submit to funding agencies. Summer salary is an allowed budget item to be included in all proposals submitted to essentially all funding agencies, unless specifically excluded, which occurs on certain limited-scope grant funding opportunities. For example, academic institutions and funding agencies will sometimes offer research assistance grant opportunities, where the intent is to assist research program development, or enhance existing research projects, by providing funds to recruit additional students from certain undergraduate or underrepresented minority groups, or to support certain travel opportunities, and other such enhancement activities. These types of grant opportunities often will explicitly exclude principal investigator salary to be included. However, the majority of grant opportunities permit and, in fact, expect to see, summer salary included in the budget.

Funding agency program managers and program directors expect the principal investigator (PI) to work on the research project, and to commit a portion of their time to performing the research. The exact time commitment is detailed in the proposal, and the allotted time is included as a salary item in the project budget. The time commitment usually consists of a specified amount of time during the academic year, and a defined time period in the summer. Once this commitment is made, it needs to be honored, and the faculty member is expected to actually work on the project for the budgeted time. That is, the funding agency expects to receive the effort that they are paying for, and for which they have provided the necessary funds. To not work on the research project for the time that has been committed is a violation of the grant agreement. This can be an issue for a faculty member that has multiple research grants from multiple funding agencies. It is possible that they may have committed more than 100% of their summer months to research activities. This situation, of course, is not possible to honor and, if this occurs, the faculty member may be required to do a budget modification and adjustment so that no more than 100% of their time is committed to their summer research activities. For example, his or her time devoted to the project can be reduced, with the funds applied to other budget items, such as funding additional student effort, increased travel, additional materials or supplies, etc. The budget modification will need to be approved, either by the home institution or by the funding agency, depending upon the magnitude of the modification. However, the best approach is to plan your research activities very carefully, and make sure your time commitment to each project that you plan is accurate and adequate to perform the proposed research. The PI's time commitment to the research project is an important factor that both the program manager or program director, and the reviewers, will consider in their final evaluation of your proposal. Your time commitment to the project needs to be realistic and indicative of serious involvement in the project.

There is another restriction for faculty members that have three months of their summer effort committed to research grants. Many faculty members will have multiple grants in progress at a given time, and they are committed to devoting a fraction of their academic year and summer period to working on each grant. If they have three months of summer effort committed to research, they are faced with some severe limitations regarding other activities in which they are permitted to engage during the summer months. That is, if they are committed to working on research for the entire three-month summer period, they are not permitted to devote time to any other activity, at least not during the normal work week hours. They cannot do any work for the college or university, engage in any community activities, do any professional consulting for pay, participate in any professional society activities, such as reviewing journal papers, or attending conferences and workshops, unless these events directly relate to their research, and, especially, they are not permitted to take personal vacation time. The latter issue is particularly important since you simply cannot take a personal vacation while you are receiving salary from the US government, or any other funding source.

By accepting three months of salary you have totally devoted your entire summer to your research activities. This creates a problem when faculty members are expected to participate during the summer months in certain university activities, such as committee work, new faculty recruitment efforts, meeting with visitors, etc., or when he or she wishes to take family vacations. Even though the university does not provide any salary during the summer months, many departments still expect their faculty members to participate in certain events and activities. In order to avoid this problem, some academic institutions limit the total amount of summer time that faculty members can commit to all research projects to no more than two months, while some will permit two and half months to be devoted to research activities. This allows time for participation in other summer activities. The exact time each faculty member spends on his or her research projects is documented and reported on an effort report that must be completed for each research grant. The time effort must be reviewed, approved, and signed by each faculty member.

The time effort reports are kept on file, and are available to US government auditors. The faculty members' total time commitment cannot exceed 100%.

8.3 Export Control Laws and Academic Research

For reasons of national security and foreign policy, the US government controls and limits the access to certain information and some specific items by foreign individuals and organizations. Federal export control laws restrict the shipment, transmission or transfer of certain items, some specific technology items and related information, as well as some types of software, and certain services, from the USA to foreign countries. There are also restrictions on what is termed "deemed exports," which involves the transfer of information related to restricted or controlled technology, software source code, and other controlled items, to foreign nationals located in the USA, including foreign students enrolled in US academic institutions. For example, instructing a student from a foreign country in the use of an instrument that is on the restricted list is considered a "deemed export." This can be a problem for research projects that are identified with export limitations, and obtained from certain funding agencies. However, even if the particular technology is not funded from one of these agencies, the technology item, if it is on the restricted list, cannot be released or transferred to any foreign entity without US government authorization, which is provided by means of an export license. Fortunately, for many university research projects, the US government has provided an "exception" that effectively permits research activities to be conducted without the need for an export license, even when the research is to involve foreign students, and the results of the research are to be presented internationally. There are two major export control laws that affect academic research: (1) ITAR (International Traffic in Arms Regulation) administered by the Directorate of Defense Trade Controls (DDTC) at the State Department, and (2) EAR (Export Administration Regulations) administered by the Bureau of Industry and Security (BIS) at the Department of Commerce (DOC). Both of these regulations are designed to help ensure that sensitive defense related information and technology do not get into the wrong hands. An export license, by which the US government grants permission to transport or sell items that fall under ITAR or EAR, must be obtained before any information or items covered under these regulations can be transferred to foreign countries, organizations, or individuals. These restrictions and the applicable procedures are explained in this section, including the academic research exclusion.

8.3.1 ITAR (International Traffic in Arms Regulations)

The ITAR regulations were designed and written to address information and technology associated with direct defense-related use and applications. The various items that are subject to ITAR are listed in the US Munitions List (USML) or the Missile Technology Control Regime (MTCR) Annex. Items listed on the USML are organized into 21 categories, and grouped into subject areas. Among the categories are areas where there is significant university research activity, such as: ground vehicles; aircraft and related articles; toxicological agents; military electronics; materials and miscellaneous articles; missile and spacecraft systems; satellite equipment and related software; and many others. The USML is periodically reviewed and the included items change over time. All items, information, and services contained on these lists are covered by the ITAR regulations. In particular, items and technology which have been identified as possessing "capacity for substantial military utility or capability," such as tanks, high explosives, naval vessels, attack helicopters, and other major equipment and system items, are identified as Significant Military Equipment (SME), and are noted on the USML with an asterisk, and are the most controlled items. Also, the USML states that "any technical data directly related to the manufacture or production of any defense articles identified as Significant Military Equipment is also designated as SME." As a practical matter, the ITAR regulations state that any and all information and material pertaining to defense and military related technologies listed on the USML can only be transferred to or shared with individuals who are either US citizens or permanent residents, unless specific authorization from the Department of State, in the form of an export license, is received, or a special exemption is applicable.

ITAR affects university research projects since certain grant funding opportunities may be classed under ITAR, or the research topic may include ITAR-related items. Participation in these programs may require prior written authorization, in the form of a license, from one or more US government agencies in order to perform sponsored research, or other educational activities involving specified technologies, or interact with people, including students or faculty members from certain countries. This can present difficulties for faculty members who submit proposals. If you submit a proposal to one of these programs, the US government funding agency will process your proposal, and you may potentially receive a grant. However, once you receive the grant you are responsible for maintaining all ITAR safeguards, and for following all appropriate policies and procedures. This could include the participation in the research project of only students, post-doctoral research assistants, or staff, who are US citizens or permanent residents. These participants would be prevented from discussing the research with any non-US citizen colleagues, including other graduate students or post-docs that may be in your research group. This is obviously very challenging if your graduate students work in a common laboratory, or share office space, with foreign students. If you receive an ITAR-classified research project, you will be required to personally sign a declaration that you agree to follow the ITAR restrictions, and you would be held personally responsible for any violation. Convictions for ITAR-related violations can involve administrative sanctions, such as the loss of research funding, or the imposition of heavy financial penalties, and possible imprisonment. For this reason, many university faculty members elect to not participate in ITAR classified programs. Also, some universities will not permit their faculty to participate in ITAR-related projects. Before submitting a proposal for a grant funding opportunity that includes ITAR classification and restrictions, you should consult with your department head or chair, or someone from your research administration office. Most colleges and universities have research administration personnel who are well versed in export regulations and restrictions, and their effects upon academic research.

Fortunately, ITAR does not apply to information generally considered to be in the public domain. This includes information related to general scientific, mathematical, or engineering principles that are commonly taught in schools and colleges and readily available in textbooks, technical publications, magazines, etc. Information that is generally available on the internet may also be exempt. ITAR does not apply to general marketing information or basic system design and operating principles and description information that is readily available publicly.

More information regarding ITAR can be found at: https://www.pmddtc.state.gov/regulations_laws/itar.html

8.3.2 EAR (Export Administration Regulations)

The EAR are also important to recognize. These regulations apply to what are called "dual use" items, consisting of commodities, software, equipment or other technology-related items that have both civilian and military uses. Items included under the EAR generally have less sensitive military application than those covered under ITAR. The items and topics that are included are listed in the Commerce Control List (CCL) included in the EAR, and are grouped into 10 categories that include many topics and areas of active university research. For example, research topics such as biotechnology, advanced materials, microelectronics, computing, telecommunications, encryption, etc., are included in the CCL. The Commerce Control List (CCL) is similar to the USML included under ITAR. The CCL specifically addresses general topics that are possible areas for university-based research, and under which many research projects will originate. For example, areas such as chemical and biological weapons technology, nuclear nonproliferation, national security, missile technology, crime control, and anti-terrorism, are identified and included in the CCL. The level of control can vary, and depends upon the country, as well as to whom or which organization within the country, the item is being transferred, as well as the intended purpose and end-use for the item. The EAR control sensitive items that: (1) have their origin in the USA, (2) are made or manufactured with US materials and technology, or (3) are not under the jurisdiction of another set of regulations, such as ITAR. The EAR establish and define the rules and regulations that a company, business, or academic institution will follow to determine whether items and activities are subject to the EAR, and to ensure that they are in compliance with export requirements.

More information regarding EAR can be found at: https://www.bis .doc.gov/index.php/regulations/export-administration-regulations-ear

8.3.3 NSDD 189 and Fundamental Research Exclusions

During the Cold War between the United States and the Soviet Union, which occurred in the late 1960s and early 1970s, there were tensions and concerns over the potential exposure of certain information deemed to be critical to national security. For this reason, the US government passed legislation that established export control regulations in an attempt to protect sensitive information and technology. Of particular concern to academic institutions and faculty researchers, there were some individuals within the government who believed that the exposure of certain information could potentially erode the scientific, engineering, and technology advantages that the nation might enjoy. They believed that the open nature of US academic institutions, where scientific and engineering research results were routinely published in the open literature and presented at scientific and engineering conferences, meetings, and workshops, could potentially result in the transfer of sensitive information and technology to unknown recipients and operators. These concerns resulted in the establishment of the export control laws, and the ITAR and EAR, in particular. The ITAR were enacted in 1976, and the EAR were enacted in 1979. These regulations were established to deny access to foreign operators, who could possibly use the information and related technology to advance the military potential of adversarial nations or to aid in the proliferation of nuclear and biological weapons, or other weapons of mass destruction. As a result, many US government funding agencies began to classify some of their academic research funding opportunities under export control regulations. There was a significant and growing expansion of research programs being announced that included the ITAR and EAR.

The academic research community became increasingly concerned over these developments. In particular, the academic research and education community was concerned about its ability to adequately comply with the export control regulations, and as a result they began communicating with US government agencies and officials in order to clarify the issue. The academic community felt that the export control restrictions were too severe, and not necessary since critical projects could be adequately protected under the established security classification procedures. There was some sympathy for this point-of-view, particularly in the research and engineering office in the US Department of Defense (DOD). In response to the concern of the academic community, officials in this office formulated a draft policy that was reviewed, and ultimately resulted in a National Security Decision Directive (NSDD), that was issued from the White House, and signed by President Ronald Reagan on September 21, 1985, as NSDD 189. A National Security Decision Directive is a form of executive order issued by the President of the United States, with the advice and analysis of the National Security Council, and the directive represents the President's national security policy and carries the "full force and effect of law."

NSDD 189 established the national policy for regulating access to science, engineering, and related technology and information that results from federally funded research performed at US academic institutions, including colleges, universities, and laboratories. NSDD 189 recognized the nature of university research, the results of which are historically freely published in scientific and technical journals, without the need to secure pre-approval from the US government or other sponsors, and presented in meetings open to all participants, without restriction or limitation. The vast majority of the research performed in academic institutions for the DOD is categorized and funded as either basic research (6.1), applied research (6.2), or advanced development (6.3), and most of the academic research is funded from the basic research (6.1) and applied research (6.2) categories. (The DOD funding categories are discussed in Chapter 5.) Together, the basic research and applied research accounts make up the DOD Science and Technology program. NSDD 189 recognized this practical matter, and defined a new category, called "Fundamental Research," as the DOD support for academic research programs that are supported from the basic research and applied research accounts, no matter the specific account. More specifically, fundamental research was defined as "basic and applied research in science and engineering, the results of which ordinarily are published and shared broadly within the scientific community, as distinguished from proprietary research and from industrial development, design, production, and product utilization, the results of which ordinarily are restricted for proprietary or national security reasons." Further, NSDD 189 stated that, to the maximum extent possible, the products of fundamental research would remain unrestricted. The classification procedure should be used when there was a matter of national security, or sensitive information that was generated during the performance of federally funded fundamental research in academic institutions. Security classification should occur before the research grant was awarded.

NSDD 189 provides an exception to the ITAR and EAR for research performed in academic institutions. This policy is commonly referred to as the "Fundamental Research Exemption," which exempts from ITAR and EAR academic research that is routinely published in the professional literature or presented at scientific and technical conferences and meetings, or other material that is readily available in the "public domain." Public domain generally refers to information that is published and is widely accessible or available to the general public.

Research activities that are considered fundamental research can generally involve foreign students and researchers, even if the research involves the design and construction of items, components, systems, or other technology intended for military applications. The US Department of State has ruled that such research activities are normal academic education and training activities, and do not constitute illegal export of a "defense service." However, the fundamental research exemption does not, in general, apply if the college or university accepts the requirement that a grant funding sponsor be permitted to review and pre-approve a publication before it is submitted to a journal or conference, regardless of whether the sponsor is federal, industry, or non-profit organization. However, universities may agree to pre-publication review in order to prevent inadvertent disclosure of sponsor proprietary information. They may also agree to a brief delay in publication to ensure that patent rights will not be lost without jeopardizing the exemption.

NSDD 189 has been reaffirmed several times by succeeding administrations since being issued in 1985, and is still in effect today. It provides the basis for academic research to be performed in subject areas and on topics that would normally be restricted according to ITAR and EAR. If you wish to submit a research grant proposal that includes ITAR or EAR restrictions, you should first consult with your department head or chair and your institution's research office, to determine the specific policies and compliance requirements that would be involved.

8.3.4 International Travel

Many university faculty members, and their students, engaged in research will often travel to conferences, workshops, or to visit colleagues at international locations. However, before planning an international trip, you need to be aware that your trip may be limited or restricted by export control regulations, particularly if you are involved in research in an area that involves sensitive information. United States export control laws are principally concerned with whether an academic researcher will travel with information or data involving controlled technology or other sensitive material that may potentially be disclosed to non-US individuals or organizations. Export regulations include items such as sensors, test instruments, reagents, biological materials, and other items, as well as data, reports, publications, etc., that may be located on your laptop computer, or other electronic device, including flash memory sticks, DVDs, smart phones, etc. Often faculty members, particularly those collaborating with or supported by industrial organizations, will have data or other information provided by the industrial source located on their laptop computer. Some of the data may be sensitive, and controlled under the ITAR or EAR regulations, and you may be unaware of the sensitivity of the information. If so, there may be certain travel restrictions that define what devices or information you are permitted to take with you. For instance, if you intend to perform some field research with a colleague in another country or to provide a technological demonstration at a conference, you may need to send to the site separately certain equipment, such as computers, sensors, measuring instruments, reagents, etc.

Before embarking upon your travel, you need to determine if you will be taking any ITAR or EAR classified items or information with you, and whether or not your trip is subject to ITAR or EAR. If so, you need to make sure that your travel is in compliance with the export control regulations. If one or more of the regulations applies to a proposed trip, then the academic traveler must determine if those regulations provide an exclusion or exemption for the type of disclosure or export item or information that will permit the proposed foreign travel and activity without getting an export license from the appropriate agency. However, it is important for you to understand that the "fundamental research" exemption does not directly apply to any locations outside the USA. That is, the fundamental research exemption that applies to academic research within the USA is not applicable when traveling abroad. Therefore, you need to protect any data, information, or other sensitive items that you may take with you, including material for presentations, lectures, etc. It's best to take a clean laptop that contains only the material you intend to present, and to make sure all computers or other electronic devices stay in your personal possession at all times. Also, it's best to not permit USB sticks, etc., to be connected to your laptop or computer. They may contain viruses or other malevolent software. Before planning your trip you should discuss export compliance policies with your research administration or travel office. The penalties for disclosure, even if inadvertent, to non-US persons or organizations of export controlled information or items can be severe.

8.4 What We've Learned

There are certain rules and regulations that must be followed when developing a funded research program. In particular, disclosure of any involvement with potential funding sources needs to be disclosed to your home institution, as well as any financial income you receive from external sources. Most importantly, if you prepare a proposal and plan to submit it to multiple funding agencies, you may do so; however, you must disclose to each funding agency the other funding agencies to which the proposal has also been submitted. We've also discussed that the US government has enacted a series of export control rules and regulations that cover selected items, technology, and information that may be transferred to foreign individuals and organizations. There is a category called "deemed exports," which includes education and training of foreign students in US academic institutions. Technology and information, which have military applications, are export controlled under the International Trade in Armaments Regulations (ITAR) administered by the Department of State, and items that have "dual-use" capability, that is the item can be used in both military and commercial applications, are export controlled under the Export Administration Regulations (EAR) administered by the Department of Commerce. We also discussed cautions associated with international travel, particularly if any item or information that is controlled under the ITAR or EAR is involved.

9   You're on Your Way

The purpose and intent of this book has been to inform and educate those interested in pursuing an academic career on the best methods for proceeding. Recognizing the importance to a faculty member of establishing an academic research program, a main objective of the book is to explain and describe how the research grant funding system functions, both from the academic institution and grant funding organization perspectives. We started the discussion with a brief history of how the US government became a major supporter of research performed in academic institutions. The US government's interests in research performed in academic institutions have historically been associated with obtaining the results of research for use in certain applications directly related to their goals. Although there has been government support for academic activities dating back to the founding of the nation, serious and significant financial assistance and support began during the Second World War. Most of the financial support from the early days was directed to research that was associated with military activities, and this support has continued up to the present day. However, since the Second World War, US government research support to academic institutions has increased, and a variety of offices and agencies throughout government have been established to manage and execute the process. Major agencies such as the Department of Energy (DOE), NASA, the National Institutes of Health (NIH), and many others now actively provide significant financial support for research performed in academic institutions. These offices and agencies consist of those that were established for the support of US government agency missions, and a non-mission agency, the National Science Foundation (NSF), which was established to support research activities in science, engineering, and education, in general. A variety of other research funding organizations has also been created and established by local and state governments, industrial and business organizations, and private foundations. These agencies all operate in similar, but fundamentally different manners. For example, US and state government agencies often will advertise research funding opportunities for specified and directed purposes. Mission agency program managers generally have an end-use in mind for the research topics they support, and in order to obtain financial support, a prospective researcher needs to learn the interests and goals of the program manager in order to successfully compete for research grants. This principle also holds for certain grant funding opportunities offered by local and state government agencies, industrial and business organizations, and some not-for-profit foundations. For these opportunities, the research proposal needs to directly address the interest area and specific objectives of the funding opportunity. Conversely, program directors at NSF will generally accept research proposals on essentially any topic, provided it is on a topic that falls within the interest area they support.

In Chapter 3 we digressed from the subject of building an externally funded research program, and took a step backwards, and discussed the academic recruitment process, and how it functions. The purpose was to address questions that someone who is interested in an academic career, but doesn't yet understand the recruitment, application, and hiring process, might pose. This material is primarily directed towards new PhD graduates, or those with some experience in industrial, business, or government organizations, who wish to transition to a position in academia. It is important to understand the academic process and, in particular, the academic faculty member performance evaluation and review process, and the promotion and tenure requirements. The main point is that once you accept an academic faculty member position, the clock starts ticking, and your performance will be reviewed on a periodic basis. Since your research performance is a critical factor in your evaluation, you need to establish a research program as soon as possible, and within your first year of academic employment. For this reason, it is very important to secure some initial funding that can be used to begin your research program. During the recruitment process, you will be able to negotiate certain resources in what is called a "start-up" package.

The items that will be of most significance were identified and discussed. Other methods for obtaining "seed" funding are also addressed

In this book we have stated and emphasized that it is fundamentally important to personally communicate with program managers or program directors in grant funding agencies, in order for them to get to know you, learn what research you are pursuing, and for them to gain confidence in you as a potential participant in the research program they support. How to identify an agency that supports your research area, and how to make contact, were addressed. Various methods for identifying, contacting, and communicating with research funding agency program officers appropriate for your research area were presented. The use of funding agency websites and, in particular, the Grants.gov website, is emphasized. We also discussed how to identify funding sources and organizations that are most likely to have the most funding opportunities. This issue, of course, relates to the budget and financial resources available to them. For this reason, we discussed the budgetary planning process for US government funding agencies. Along this line, the various legal instruments the government uses to transfer funds from the government to external providers of products and services are defined and described. The most commonly employed instrument is the "grant," which is basically a "gift," although certain requirements and conditions usually accompany the award.

Research proposals all have a defined and specific structure. The basic elements of a research proposal were presented and discussed, in some detail, in Chapter 7. I started the chapter with some general comments from my experience in reading, reviewing, and evaluating a large number of proposals over an extended period of time. I noted how an experienced proposal reviewer will read and evaluate proposals, and the need for the proposal author to include all requested information, but to be concise and clear in the information they present. You don't want reviewers to need to hunt for important details. This information was extended to four fundamental principles that should be recognized and followed when preparing a proposal. The elements of a basic proposal were described, and how to address each of these elements was discussed. The NSF CAREER proposal is fundamentally different from a regular research grant proposal in that the CAREER proposal addresses a life-long career plan, and therefore, needs to include both a research and an education plan. In particular, the integration of the research and education programs needs to be described. Details of the CAREER proposal were separately discussed, owing to its importance to new faculty members. The CAREER award has become extremely important and is very highly regarded by both academic institutions, as well as grant funding agencies. The award of a CAREER grant is considered a sign of quality performance, and can be very important in a faculty member's evaluations for promotion and the award of permanent tenure.

The book concludes with a discussion of some important limitations and restrictions associated with export control by the US government. In particular, research associated with certain items, topics, and related information may be identified as "sensitive," and may be export controlled under ITAR and EAR. If so, special precautions need to be followed and respected. Included is training and education to individuals or organization from specified foreign countries, which is called "deemed export." Fortunately, research termed "fundamental research" may be exempted from the export controlled regulations. These issues are described in Chapter 8.

It is my sincere hope that the information presented in this book is helpful to those in the process of initiating an academic career. I've presented information I've learned while working in academic, industrial, and US government organizations in a career extending over four decades. I've presented the information, where possible, from the perspective of a US government funding agency program manager and program director in an attempt to give the reader an idea of how they review proposals, and what information they expect to see. You want to make it easy for them to find information that is specifically requested, and to make their job of reading your proposal as easy as possible. If you do this, the rewards will be great. While no one can predict the future, if you follow the guidelines presented in this book, you have a good chance for success. You now know the rules, and the rest is up to you. I wish you success in your efforts. Welcome to your new occupation and good luck in obtaining grant funding and building your academic research program.

Index